MIX
Papier aus verantwortungsvollen Quellen
Paper from responsible sources
FSC® C105338

Dr. Omed Abdullah

Modeling the Lattice Parameters of Solid Solution Alloys

Anchor Academic
Publishing

Abdullah, Omed: Modeling the Lattice Parameters of Solid Solution Alloys, Hamburg, Anchor Academic Publishing 2017

Buch-ISBN: 978-3-96067-098-8
PDF-eBook-ISBN: 978-3-96067-598-3
Druck/Herstellung: Anchor Academic Publishing, Hamburg, 2017

Bibliografische Information der Deutschen Nationalbibliothek:
Die Deutsche Nationalbibliothek verzeichnet diese Publikation in der Deutschen Nationalbibliografie; detaillierte bibliografische Daten sind im Internet über http://dnb.d-nb.de abrufbar.

Bibliographical Information of the German National Library:
The German National Library lists this publication in the German National Bibliography. Detailed bibliographic data can be found at: http://dnb.d-nb.de

All rights reserved. This publication may not be reproduced, stored in a retrieval system or transmitted, in any form or by any means, electronic, mechanical, photocopying, recording or otherwise, without the prior permission of the publishers.

Das Werk einschließlich aller seiner Teile ist urheberrechtlich geschützt. Jede Verwertung außerhalb der Grenzen des Urheberrechtsgesetzes ist ohne Zustimmung des Verlages unzulässig und strafbar. Dies gilt insbesondere für Vervielfältigungen, Übersetzungen, Mikroverfilmungen und die Einspeicherung und Bearbeitung in elektronischen Systemen.

Die Wiedergabe von Gebrauchsnamen, Handelsnamen, Warenbezeichnungen usw. in diesem Werk berechtigt auch ohne besondere Kennzeichnung nicht zu der Annahme, dass solche Namen im Sinne der Warenzeichen- und Markenschutz-Gesetzgebung als frei zu betrachten wären und daher von jedermann benutzt werden dürften.

Die Informationen in diesem Werk wurden mit Sorgfalt erarbeitet. Dennoch können Fehler nicht vollständig ausgeschlossen werden und die Diplomica Verlag GmbH, die Autoren oder Übersetzer übernehmen keine juristische Verantwortung oder irgendeine Haftung für evtl. verbliebene fehlerhafte Angaben und deren Folgen.

Alle Rechte vorbehalten

© Anchor Academic Publishing, Imprint der Diplomica Verlag GmbH
Hermannstal 119k, 22119 Hamburg
http://www.diplomica-verlag.de, Hamburg 2017
Printed in Germany

SUMMARY

Models for the prediction of lattice parameters of substitutional and interstitial solid solutions as a function of concentration and temperature are presented. For substitutional solid solution, the method is based on the hypothesis that the measured lattice parameter versus concentration is the average of the interatomic spacing within a selected region of a Bravais lattice. The model which is applied on Ni-Cu and Ge-Si solid solutions gave good agreement with the experimental data.

For the interstitial solid solution of Fe-C system, the method is based on the assumption that the change in lattice parameter of the pure Fe phase is due to the occupation by carbon atoms to the octahedral holes in the *fcc* austenite; and *bct* martensite. Results of this model for the Fe-C martensite are in a good agreement with experimental data, however the preparation procedure of the Fe-C austenite at room temperature, causes crystal defects thereby dropping the experimental values by 0.25% from the theoretical ones.

The model of lattice parameter versus temperature for both substitutional and interstitial solid solutions is based on the relative change in length and vacancy concentration at lattice sites that are in thermal equilibrium. Combination of both models then facilitate the calculation of lattice parameters as a function of concentration and temperature. Results are discussed accordingly.

CONTENTS

CHAPTER ONE
Literature outline of basic concepts

1-1 Introduction 1
1-2 Solid solutions 1
      1-2-1 Substituational solid solution 2
      1-2-2 Interstitial solid solution 4
1-3 Atomic radius and coordination number 5
1-4 Modeling of lattice parameter in solid solutions 6
      1-4-1 Vegard's law 6
      1-4-2 Departures from Vegard's law 7
      1-4-3 Moreen's model 9
      1-4-4 Models on Fe-C system 10
1-5 Point defects 11
1-6 Theme of work 12

CHAPTER TWO
Theoretical aspects

2-1 Introduction 13
2-2 Relation between atomic and weight percentage 15
2-3 Lattice parameters of substitutional solid solution 16
2-4 Lattice parameters of interstitial solid solution 18
      2-4-1 Austenitic Fe-C system: face centered cubic case 18
      2-4-2 Martensitic Fe-C system: body centered tetragonal case 19
      2-4-3 Ferrite Fe-C system: body centered cubic case 22
2-5 Lattice parameters versus temperature changes 23
2-6 Lattice parameters versus concentration and temperature 24
2-7 Computer programming 25

CHAPTER THREE
Results and Discussion

3-1 Lattice parameters versus concentration 28
      3-1-1 The Ni-Cu system 28
      3-1-2 The Ge-Si system 31
      3-1-3 The Fe-C system 33

3-2 Lattice parameters versus temperature 37
    3-2-1 The Ni-Cu system... 37
    3-2-2 The Fe-C system.. 42

CHAPTER FOUR
Conclusions and Suggestion for future work

4-1 Conclusions.. 50
4-2 Suggestions for future work.. 51

REFERENCES 52

APPENDICES
Appendix A: Listing of the programme LAT PAR 57
Appendix B: Phase diagrams 60
Appendix C: Glossary of terms related to physical metallurgy
              used in this study 62

LIST OF SYMBOLS

a	: lattice parameter.
a_o	: original lattice parameter.
$a^{(ave)}$	: average lattice parameter of the array.
B	: bulk modulus.
c	: lattice parameter of bct.
C	: atomic concentration.
C_i	: atomic fraction of i species.
C_j	: atomic fraction of j species.
d	: near-neighbor interatomic distance.
E_i	: energy required to form an interstitial defects.
E_v	: energy required to form an vacancy defects.
k	: Boltzman constant.
n	: number of Frenkel defects.
N	: number of atoms.
N_i	: number of interstitial.
N_v	: number of vacancies.
P	: Probability.
r	: atomic radius.
$S^{(ave)}$	: average interatomic spacing of the array.
$Si^{(ave)}$	: average spacing about an atom of the i species.
S_{ij}	: interatomic spacing between atoms of i and j species.
T	: absolute temperature.
V	: atomic volume.
x_c	: number of carbon atoms per 100 iron atoms.
α	: thermal expansion coefficient.
μ	: shear modules.
Δ	: departures from Vegard's law.
Δa	: change of lattice parameters.
Δc	: change of c-axis of bct.
ΔL	: change of length.
ΔT	: change in temperature.

CHAPTER ONE
Literature outline of basic concepts

1-1 Introduction

An alloy is a substance that has metallic properties and is composed of two or more chemical elements, of which at least one is a metal. There are two possible phases in an alloy, intermediate alloy phase or compound, and solid solution. If an alloy is homogeneous (composed of a single phase) in the solid state, it can be only a solid solution or a compound. If the alloy is a mixture, it is then composed of any combination of the phases possible in the alloy [1].

One important characteristic of a metals alloy is its lattice parameters, which can be measured by observing the diffraction of either X-rays, neutrons, or electrons [2], or can be predicted on the basis of the crystal structure of its constituents and their concentrations [3], can be utilized in many practical applications.

1-2 Solid Solutions

A characteristic property of metals is that if two (or more) are melted together in suitable proportions a homogeneous solution often results. When cooled this is called a solid solution because, as in the case of a liquid solution, the solute and solvent atoms are arranged at random [3]. Random arrangement of the two kinds of metal atom is always found if the alloy is cooled rapidly (quenched). In certain solid solution with particular concentrations of solute a regular atomic arrangement develops on slow cooling [4].

1-2-1 Substitutional Solid Solution

The substitutional solid solution is the more general case, that the solute atoms replace those of the solvent, so that the two kinds of atom are situated on a common lattice. The substitutional solid solutions may be accompanied by either an increase or a decrease in cell volume, depending on whether the solute atom is larger or smaller than the solvent atom [5].

Substitutional solid solutions are of two types:
  (i)   Random substitution solid solutions.
  (ii) Ordered substitutional solid solutions.

When there is no order in the substitution of the two elements (as shown in Fig. (1-1a)), the chance of one element occupying any particular atomic site in the crystal is equal to the atomic percent of that element in the alloy. In such a case the concentration of solute atoms can vary considerably throughout the lattice structure. The resulting solid solution is called a random or disordered substitutional solid solution. If the atoms of the solute material occupy similar lattice points within the crystal structure of the solvent material, this is called an ordered solution (Fig. (1-1b)). Such ordering is common at lower temperatures since greater thermal agitation tends to destroy the orderly arrangement [6].

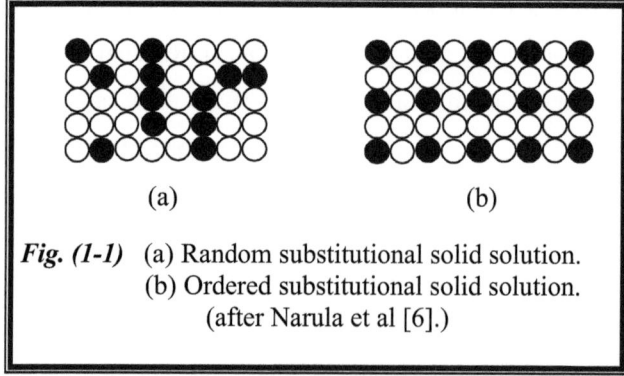

Fig. (1-1) (a) Random substitutional solid solution.
              (b) Ordered substitutional solid solution.
                  (after Narula et al [6].)

In general, the substitutional solid solutions are not formed by all pairs of metals due to principal physical factors that control the range of solubility in solid solutions which are:

i- *Atomic Size Factor:*
Hume-Rothery et al [3] advanced the hypothesis that the size factor is favorable for solid solution formation when the difference in atomic radii is less than about 15 percent. If the atomic size factor is greater than 8 percent but less than 15 percent, the alloy system usually shows a minimum solubility. If the atomic size factor is greater than 15 percent, solid solution formation is very limited.

ii- *Crystal Structure Factor:*
Continuous solid solubility is possible for two metals that have identical crystal structures, except for the dimensions of the unit cell which are governed by the atomic size factor. Thus continuous solid solutions are possible between face centered cubic Cu-Ni or body centered cubic Mo-W, but not between body centered cubic Mo and face centered cubic Cu [1].

iii- *Electronegativity Factor:*
If the two kinds of atoms in a solid solution are respectively electronegative and electropositive, then it is likely that they prefer to form stable structure rather than continuous solid solutions [1]. Because these structures frequently have compositional variations over a certain range, it is proper to call such crystals intermediate phases rather than compounds [3].

iv- *Relative Valency Factor:*
Continuous solid solutions can occur only between atoms having the same valency in the alloy. It is generally true that elements of lower valency dissolve to a larger extent in a higher valency solvents than the reverse case [7].

1-2-2 Interstitial solid solution

The interstitial solid solution is the more restricted case that the solute atoms fit into the spaces between those of the solvent, so that interstitial type of solid solution is confined to cases where one atom is very much smaller than the other (see Fig. (1-2)), and the most important interstitial solutes are carbon, nitrogen, hydrogen, and boron [4].

The interstitial addition is always accompanied by an increase in the volume of the unit cell. If the solvent structure is cubic then the single lattice parameter a must increase, but if it is not cubic, then one parameter may increase and the other decrease, as long as these changes result in an increase in cell volume [3,5].

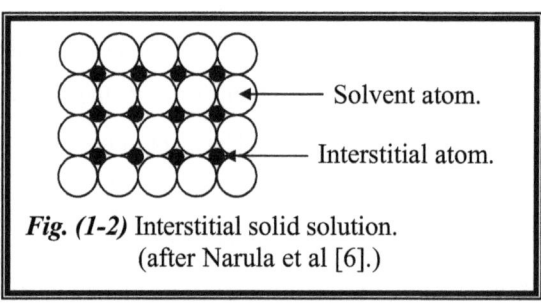

Fig. (1-2) Interstitial solid solution.
(after Narula et al [6].)

Atoms of radii less than one tenth of a nanometer can squeeze into the interstitial sites in most metals and according to the argument proposed in Ref. [8] that if the ratio of the atomic radius of the solute atom, to that of the solvent, was less than 0.59 a situation favour the formation of interstitial solid solution (see Fig. (1-3)).

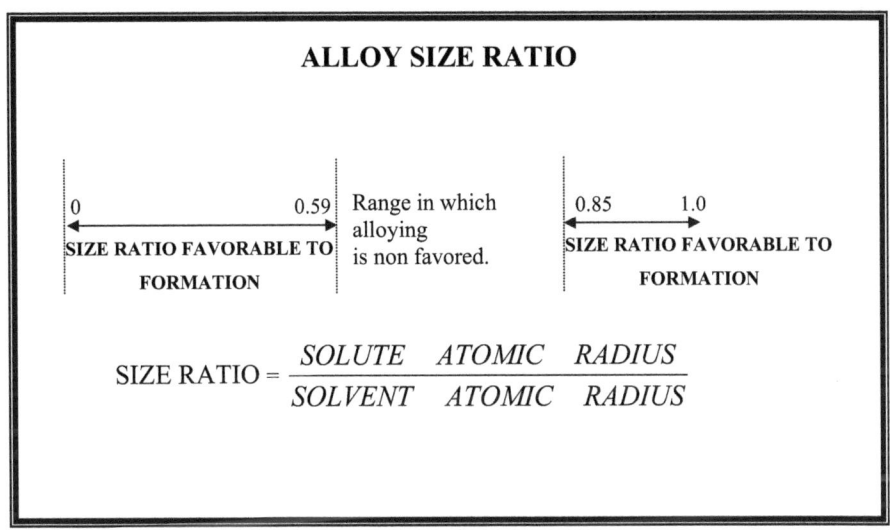

Fig. (1-3) Atom size rules for substitutional and interstitial alloys. (after Seeger et al. [8]).

1-3 Atomic radius and coordination number

Atomic radius is defined as half the distance between the nearest neighbours in the crystal structure of a pure element [6]. The size of an atom is affected by its coordination number. The type of forces existing between the atoms also affects it, and these forces may differ in unlike compounds or even different structural arrangements of the same compound. It is important, therefore, to realize that absolute values of atomic radii are only approximately known. Furthermore, the same atom can have several different radii according to its coordination number and the type of compound in which it occur [1,9].

The atomic radii derived from interatomic distances in the structure of the elements are structure-dependent parameters, whereas atomic volumes are much more nearly independent of structure, and this applies even in the face centered cubic and body centered cubic structures. For example, by equating the atomic volumes V of an element in the fcc (CN12) and bcc (CN8) structures:

$$V = a_{fcc}^3 / 4 = a_{bcc}^3 / 2$$

It is seen that the near-neighbor interatomic distance d, is about 3% larger in the fcc structure:

$$\frac{d_{fcc}}{d_{bcc}} = \frac{a_{fcc}/\sqrt{2}}{\sqrt{3}\, a_{bcc}/2} = \frac{\sqrt[3]{2}\, a_{bcc}/\sqrt{2}}{\sqrt{3}\, a_{bcc}/2} = 1.03$$

And empirically it is indeed found to be some $2\frac{1}{2}$ to 3% larger in fcc structures, implying that atomic volumes in the two structures are nearly equal [10].

1-4 Modeling of lattice parameter in solid solutions
1-4-1 Vegard's law
Vegard's law calls for a linear variation of the lattice parameters a, as a function of atomic concentration C, in substitutional ionic solid solution between two components A and B of similar structure [10].

$$a = C_A a_A + C_B a_B \qquad (1.1)$$

where a_A and a_B are the lattice parameters of solvent and solute respectively.

Vegard's law really applies to solid solutions of salts [11,12], but it did not predict the lattice parameters correctly in accordance with the experimental data for metallic solid solutions [10]. This law is proposed by Vegard in (1921), since then several papers have been written concerning departures from Vegard's law and means for corrections to match with the experimentation.

The behavior can be shown in Fig. (1-4), the actual lattice parameter in curve (a) lies above that for linear relation, there is said to be a positive deviation from Vegard's law, whereas a curve such as that of (b) corresponds to a negative deviation [3].

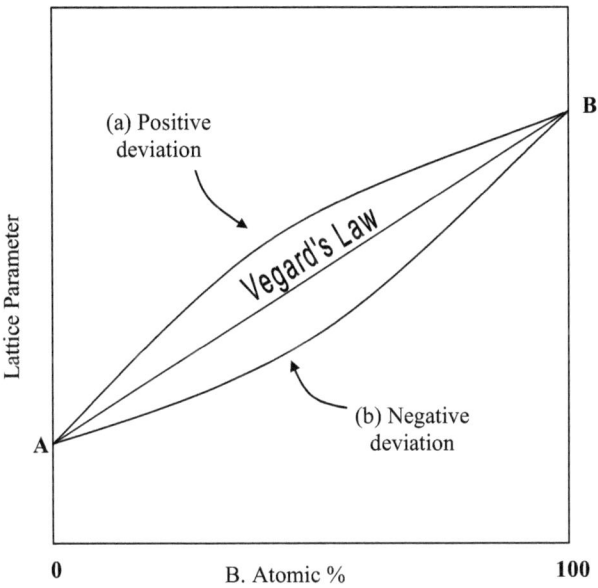

Fig. (1-4) Illustration of positive and negative deviation from Vegard's Law. (after Hume-Rothery et al.[3]).

1-4-2 Departures from Vegard's law

Numerous attempts have been made to calculate and predict departure from Vegard's law. Some of the models will be mentioned below.

Jaswon, Henry, and Raynor [13] investigated copper, silver and gold alloys in an attempt to explain departures from Vegard's law based on strain energy due to the introduction of a solute into the matrix. Their analysis lead to the following equation:

$$\Delta = C_B [13(a - a_A) - (a_B - a_A)] \qquad (1.2)$$

7

where Δ is the departures from Vegard's law, the subscript A refers to solvent and B to the solute, C is the atomic concentration, a_A and a_B are the lattice parameters of solvent and solute respectively, and a (without a subscript) refers to the lattice parameter of the alloy.

Zen [14] showed that Vegard's law is valid only when the volumes of the two components are approximately equal. He suggested that if the specific volumes of the two components are significantly different, then Vegard's law does not hold, because it is the volumes rather than the lattice parameters of the two components which are additive.

Zen suggests that Vegard's law should be expressed as:

$$a = a_A \left[1 - \left\{ 1 - (a_B / a_A)^3 \right\} C_B \right]^{1/3} \qquad (1.3)$$

definition of symbols as in equation (1.2).

Dienes [15] has derived a relation between the nearest-neighbor distance, the composition and the short-range order parameter for binary alloys Cu-Au and Au-Ag. The model shows that ordering has some influence on departures from Vegard's law.

d'Heurle, Nowick, and Seraphim [16] have proposed another model to calculate the lattice parameter, with or without short-range order, involving a concept of preferred bond distances.

Gschneidner and Vineyard [17] applied second-order elasticity theory to obtain the equation:

$$\Delta = 2 \left(\frac{d\mu}{dp} - \frac{\mu}{B} \right) \frac{(a_A - a_B)^2}{a_A} C_B \qquad (1.4)$$

where μ is the shear modulus, P is the pressure, and B is the bulk modulus. This equation is valid only for dilute solutions, i.e. when $C_B \ll 1$.

The magnitude and sign of the departures from Vegard's law, is poorly predicted for all models [10]. This suggests that the factor(s) mainly responsible for departures from the law have not yet been considered, and that the above considerations are only secondary influences which are general to the problem, giving results that are essentially independent of the various models [10].

1-4-3 Moreen's model

In 1971 Moreen et al [18] presented a model to predict the lattice parameters of metallic solid solutions as a function of composition, this method is based on the hypothesis that the measured lattice parameter of a solid solution alloy is the average of all the interatomic spacing within a selected region of the lattice. Except little deviation in some systems, there is a good agreement between the calculated and experimental values.

In 1985 Ning Yuatao and Xu. Hua. [19] used Miedema's theory of electric charge shift, to modify the Moreen's model of prediction of the lattice parameters in solid solutions. The lattice parameters of Cu-Al, Cu-Si and Ni-Al solid solutions at equilibrium and Ag-Sn, Ag-La and Ag-Gd extended solid solutions were calculated and found to be in better agreement with the experimental values.

1-4-4 Models on Fe-C system

Numerous studies [20-23] on the variation of lattice parameters of Fe-C solid solution with carbon concentration have been done experimentally. It has been shown that in austenite, which is an interstitial solid solution of carbon in face centered cubic γ-iron, the addition of carbon increases the cell edge a; but in martensite which is a metastable interstitial solid solution of carbon in α-iron, the c parameter of the body centered tetragonal cell increases while the a parameter decreases, when carbon is added.

In (1990), Liu Cheng et al [24] used least-squares fitting through literature data; in the case of iron-carbon austenite yields:

$$a(x_c) = a_o + kx_c \quad (1.5)$$

where a_o is the γ-iron lattice parameter, k is a constant, and x_c is the number of carbon atoms per 100 iron atoms.

In the case of iron-carbon martensite, they yield the following relationship:

$$a(x_c) = a_o - k_1 x_c \quad (1.6)$$

$$c(x_c) = a_o + k_2 x_c \quad (1.7)$$

where a_o in this case is the α-iron lattice parameter, and k_1, k_2 are constants.

In (1995) Onink [25] during an experimental work reported that, the observed dependencies of the lattice parameters of austenite on carbon concentration and temperature can be combined and expressed as:

$$a(x_c, T) = (a_o + k\ x_c)[1 + (k_3 + k_4 x_c)T] \quad (1.8)$$

where k_3 and k_4 are constants, and T is absolute temperature. In addition, the average lattice parameter of ferrite via temperature is given by:

$$a(T) = a_o(1 + k_5 T) \quad (1.9)$$

where k_5 is a constant.

1-5 Point defects

The mathematically perfect crystal is an exceedingly useful concept. In actual crystals, however, imperfections or defects are always present and their nature and effects are often very important in understanding the properties of crystals. Ordinary materials, however contain imperfections of various kinds that produce both useful properties and also such undesirable effects as causing the strength to decrease below that of a perfect crystal [26].

The simplest imperfection is a lattice vacancy, which is a missing atom or ion, also known a Schottky defect, (Fig. (1-5a)).

The probability (P) that a given lattice site is vacant is proportional to the Boltzmann factor for thermal equilibrium: $P = exp(-E_v / kT)$, where E_v is the energy required to take an atom from a lattice site inside the crystal to a lattice site on the surface, k Boltzmann constant and T absolute temperature [27].

Another vacancy defect is the Frenkel defect in which an atom is transferred from a lattice site to an interstitial position (Fig. (1-5b)), a position not normally occupied by an atom [9].

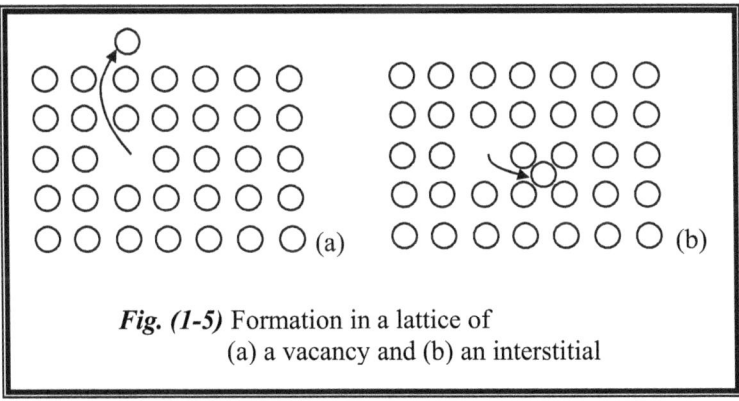

Fig. (1-5) Formation in a lattice of
(a) a vacancy and (b) an interstitial

1-6 Theme of work

Phase diagrams and/or properties of solid solutions are of great importance in metallurgy and their determination are extensively done experimentally by X-ray and thermal analysis methods.

As seen from literature outline, theoretical prediction of phase diagrams via lattice parameters received less attention, only few publications on the substitutional solid solutions and practically no available literature on the interstitial solid solution such as the common Fe-C and Fe-N. Thus the theme of work will highlight prediction of models involved in the calculation of lattice parameters via atomic radii of the binary elements and their crystal structures for the substitutional solid solution Ni-Cu and Ge-Si, and the interstitial solid solution Fe-C, subject to the correlation with concentration of solvent/solute, and in addition the model is also correlated with the vacancy defects and thermal expansion that are highly affected by temperature.

CHAPTER TWO
Theoretical Aspects

2-1 Introduction

As metals have relatively high densities, they must consist of atoms that are packed very closely together, so, for a first approximation it is permissible to consider the atoms of a metal as hard spheres packed together [26]. But it is found, that the atoms of a given metal may not always appear to have the same diameter. Nevertheless the concept of atomic diameter has proved to be a useful one in metallurgy and plays an important role in understanding the formation of alloys [28].

For the hard sphere model of the atoms, the nearest nighbour distance in a crystal of pure element is $2r$ where r is the radius of the atom; and according to this model the lattice parameter ($a_\circ$) of a crystal structure (see Fig. (2-1)) are given by the expressions [26]:

$$\left. \begin{array}{ll} bcc\ldots\ldots & a_\circ = \dfrac{4r}{\sqrt{3}} \\ fcc\ldots\ldots & a_\circ = \dfrac{4r}{\sqrt{2}} \\ Diamond\ldots & a_\circ = \dfrac{8r}{\sqrt{3}} \end{array} \right\} \quad (2.1)$$

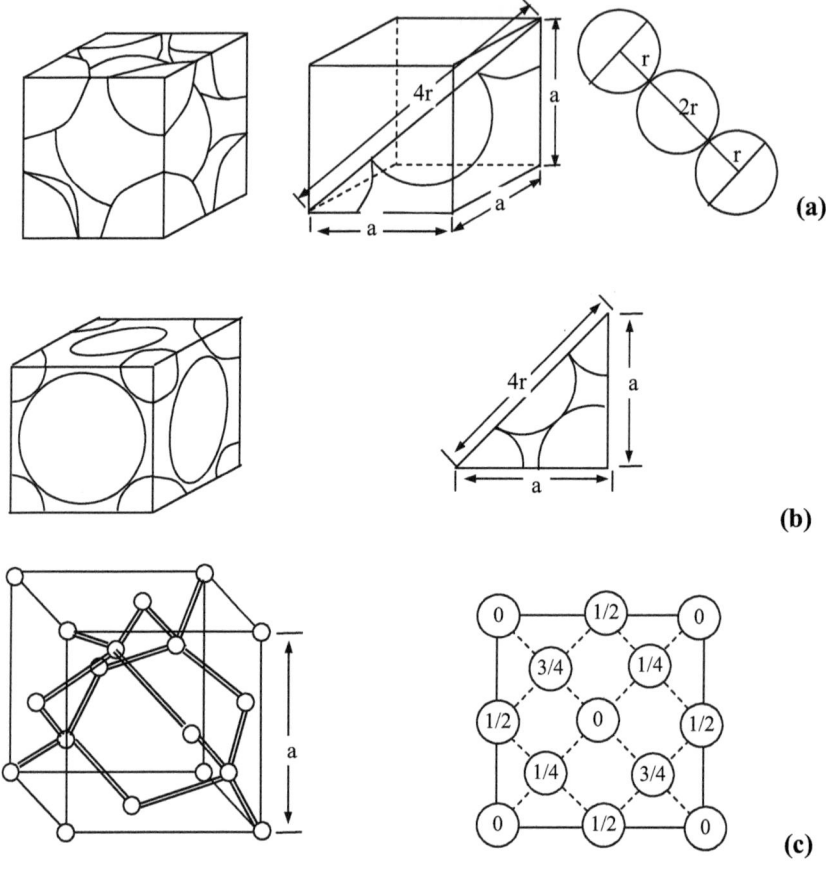

Fig. (2-1) Some metallic crystal Structures.
    (a) - Body centered cubic structure.
    (b) - Face centered cubic structure.
    (c) - Diamond structure (after Pillai[26]).

2-2 Relation between atomic and weight percentage

It is sometimes more convenient for certain types of scientific work to express the alloy composition in atomic percent. The conversion from weight percent (wt. %) to atomic percent (at. %) may be made by the following formulae [7].

$$\text{Atomic percent of } A = \frac{100X}{X + Y(\frac{M_A}{M_B})} \quad (2.2)$$

$$\text{Atomic percent of } B = \frac{100Y(\frac{M_A}{M_B})}{X + Y(\frac{M_A}{M_B})} \quad (2.3)$$

where M_A = atomic weight of A.

M_B = atomic weight of B.

X = weight percent of A.

Y = weight percent of B.

It is also convenient for interstitial alloys, such as Fe-C system, to represent the carbon concentration as the number of C atoms per 100 Fe atoms. The conversion can be made by:

$$\frac{\text{No. of C atoms}}{100 \text{ Fe atoms}} = \frac{X_c}{X_{Fe}} \quad (2.4)$$

Where X_c carbon concentration in at. %.

X_{Fe} iron concentration in at. %.

2-3 Lattice Parameters of Substitutional Solid Solution

According to the Moreen's model [18], the interatomic spacing of a random solid solution may be considered as the average interatomic spacing of all the atoms in the crystalline array, which gives:

$$S^{(ave)} = \sum_{i=1}^{n} C_i S_i^{ave} \qquad (2.5)$$

where $S^{(ave)}$ is the average interatomic spacing of the array, S_i^{ave} is the average spacing about an atom of the i species, and C_i is the atomic fraction of the i species. The probability of encountering a particular atom in a random distribution is equal to its atomic fraction. Then, counting the probabilities of encountering each type of atom that is adjacent to an atom of the i species:

$$S_i^{ave} = \sum_{j=1}^{n} S_{ij} C_j \qquad (2.6)$$

where S_{ij} is the interatomic spacing between atoms of the i and j species, and C_j is the atomic fraction of the j species. Substituting Eq. (2.6) into Eq. (2.5):

$$S^{(ave)} = \sum_{i=1}^{n} \sum_{j=1}^{n} C_i C_j S_{ij} \qquad (2.7)$$

For a binary solid solution:

$$S^{(ave)} = C_1^2 S_{11} + 2C_1 C_2 S_{12} + C_2^2 S_{22} \qquad (2.8)$$

where C_1 and C_2 are the atomic fractions of components *1* and *2*; S_{11} is the interatomic spacing between the two adjacent atoms of component *1*; S_{22} is the spacing between two adjacent atoms of component *2*; S_{12} is the spacing between an atom of component *1* and an atom of component *2*.

Eq. (2.8) describes the variation of $S^{(ave)}$ as the atomic fraction of solute changes in a binary alloy solid solution.

The separation of like atoms S_{11} or S_{22} can be found from atomic radius. However, the S_{12} term of unlike atoms was determined from experimental data by differentiating Eq. (2.8) with respect to C_2:

$$\frac{d}{dC_2}(S^{(ave)}) = -2S_{11} + 2C_2 S_{11} + 2S_{12} - 4C_2 S_{12} + 2C_2 S_{22} \quad (2.9)$$

as $C_2 \to 0$

$$\frac{d}{dC_2}(S^{(ave)})_{C_2=0} = 2S_{12} - 2S_{11} \quad (2.10)$$

Hence:

$$S_{12} = S_{11} + \frac{1}{2}\frac{d}{dC_2}(S^{(ave)})_{C_2=0} \quad (2.11)$$

Thus, the value of S_{12} becomes a simple function of the initial slope of the curve that represents the interatomic spacing of the solid solution on the basis of solute concentration.

Instead of Equation (2.8) appears in terms of the interatomic spacing, it can however be written in terms of lattice parameter $a^{(ave)}$ as:

$$a^{(ave)} = \sqrt{2}\, S^{(ave)} \quad \text{for fcc structure}$$

Therefore, equation (2.8) becomes

$$a^{(ave)} = C_1^2 a_{11} + 2C_1 C_2 a_{12} + C_2^2 a_{22} \quad (2.12)$$

The values of a_{11} and a_{22} can be found from atomic radius by using equations (2.1) according to the crystal structure of the element.

2-4 Lattice Parameters of Interstitial Solid Solution

Interstitial solid solutions normally have very limited solubility and generally are of little importance. Carbon in iron is a notable exception and forms the basis for hardening steel [7]. The lattice parameters of the various kinds of "iron" and steels, which make these materials so valuable, are dependent on the amount of carbon and the way in which it is distributed throughout the metal [29].

2-4-1 Austenitic Fe-C system: face centered cubic case

In the face centered cubic structure, the largest holes are at the centers of the cube edges and at cube centers (Fig. (2-2a)); they are surrounded by six metal atoms which form the corners of an octahedron and are thus known as the "octahedral holes". The structure also contains a second set of holes at positions such as those in Fig. (2-2b). These holes are surrounded by four metal atoms arranged tetrahedrally, and known by the "tetrahedral holes" [3].

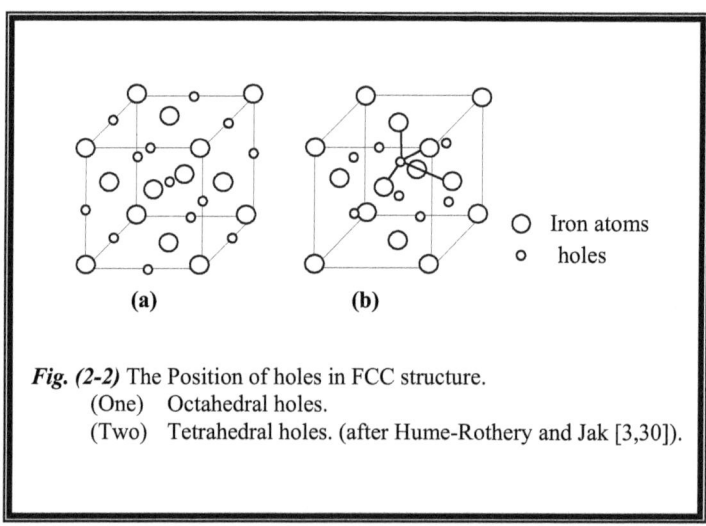

Fig. (2-2) The Position of holes in FCC structure.
  (One) Octahedral holes.
  (Two) Tetrahedral holes. (after Hume-Rothery and Jak [3,30]).

From Fig. (2-2a), if rfe is the radius of iron atoms, the octahedral holes will just contain a spheres of radius (0.414 r_{fe}). When the radius of carbon atoms is slightly greater than this value, their introduction into the octahedral holes may slightly distort the cubic structure into some lower symmetry, particularly if the number of carbon atoms is insufficient to fill all the octahedral holes [3].

In order to calculate the lattice parameters, one assume that all the octahedral holes are occupied by carbon atoms, in this case the lattice parameter is increased by the value [31]:

$$\Delta a = [2(r_{fe} + r_c) - 2\sqrt{2}\, r_{fe}] \qquad (2.13)$$

where r_c atomic radius of carbon.

Based on the maximum solubility of carbon in γFe, as only 10 at %[32], and there are four unit cells that contribute in an individual octahedron, the lattice parameter of the austenite phase in terms of carbon content becomes:

$$a = a_o + (\Delta a / 4 \times 10).X_c$$

or

$$a = 2\sqrt{2}\, r_{fe} + [\{2(r_{fe} + r_c) - 2\sqrt{2}\, r_{fe}\}/4 \times 10]X_c \qquad (2.14)$$

where a_o is the original lattice parameter, and X_c is the number of carbon atoms per 100 iron atoms.

2-4-2 Martensitic Fe-C system: body centered tetragonal case

Martensite is one of the many metastable phase of the Fe-C system. When austenite is cooled very rapidly, it transform to martensite by means of the so-called martensite transformation [1,33]. So that the martensite is an interstitial supersaturated solid solution of carbon in iron having a body centered tetragonal lattice (bct) with the distortion dipoles aligned along one expanded axis, the tetragonal c-axis [7], as shown in Fig. (2-3).

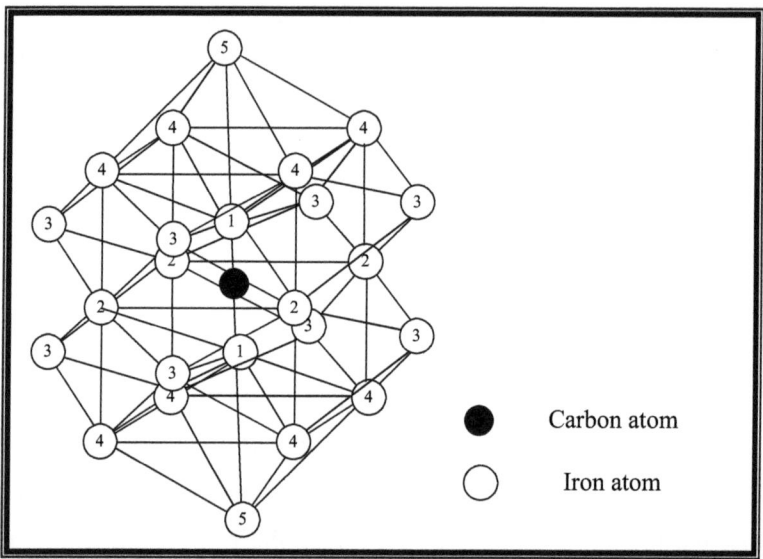

Fig. (2-3) Crystal structure of bct, iron-carbon martensite around a carbon atom. Numbers indicate near neighbor shells. (after Seeger [8]).

The actual structure can be represented by a diagram such as that in Fig. (2-4), where the elongated symbols indicate the variations in position in the vertical direction. If it is assumed that all possible positions in the unit cell are occupied by carbon, in this case the c-axis is increased by the value:

$$\Delta c = 2(r_{fe} + r_c) - 4r_{fe}/\sqrt{3} \tag{2.15}$$

and the a-axis is decreased by the value:

$$\Delta a = 4r_{fe}/\sqrt{3} - \sqrt{2}(r_{fe} + r_c) \tag{2.16}$$

by the same way as in the austenite case the lattice parameter as a function of carbon content become:

$$c = 4r_{fe}/\sqrt{3} + [\{2(r_{fe} + r_c) - 4r_{fe}/\sqrt{3}\}/(4 \times 10)]X_c \tag{2.17}$$

$$a = 4r_{fe}/\sqrt{3} - [\{4r_{fe}/\sqrt{3} - \sqrt{2}(r_{fe} + r_c)\}/(4 \times 10)]X_c \tag{2.18}$$

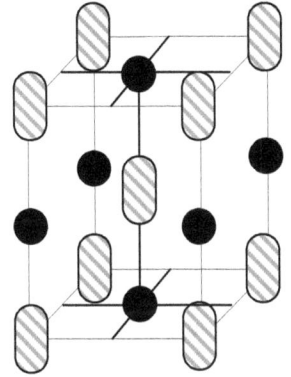

⬚ Range of Positions for iron atoms.

● Sites for Carbon atoms.

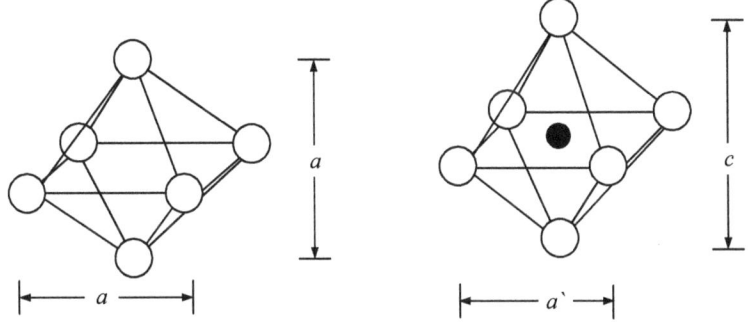

Unfilled Octahedron in α-iron   **FILLED OCTAHEDRON IN**

Fig. (2-4) The structure of martensite. (after Jack etal [30]).

2-4-3 Ferrite Fe-C system: Body centered cubic case

A very dilute solid solution of carbon in a body centered cubic iron is called ferrite. On slow cooling, austenite transforms into a mixture of ferrite and cementite [3].

As shown in Fig.(2-5) the tetrahedral holes can accommodate spheres of radius (0.291 r_{fe}). Since the carbon atoms is too big to fit comfortably in to its interstitial hole, its neighboring atoms will be displaced out word, vertically at the A sits or horizontally at the B site (Fig.(2-5)).

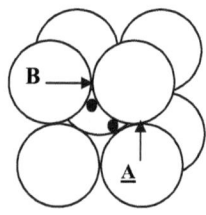

Fig.(2-5) Interstitial of the bcc, two of the four possible position on one face shown.(after Avener [7])

Therefore the distortion of the ferrite lattice by the carbon atom is much greater than in the case of austentite, but because of the much more restrict solubility of carbon, the lattice parameter of ferrite should not depend on the carbon concentration of the alloy [25].

2-5 Lattice parameters versus temperature changes

Real crystal contains intrinsic point defects in thermal equilibrium, and their concentrations are determined by the Boltzmann factor *exp(-E/kT)*. Consequently the point defect concentration increases rapidly with increasing temperature.

Statistical mechanics of crystals predicts that the number of vacancies (N_v) should depend on temperature through an expression of the form:

$$\frac{N_v}{N - N_v} = exp(-E_v / kT) \qquad (2.19)$$

where N the number of atoms, and E_v the energy required to take an atom from a lattice site inside the crystal to a lattice site on the surface [2]. If $N_v \ll N$, then:

$$N_v = N \, exp(-E_v / kT) \qquad (2.20)$$

In the same way, if the number n of Frenkel defects is much smaller than the number of lattice sites N and the number of interstitial sites N_i, the expression is:

$$n \cong (N \, N_i)^{1/2} \, exp(-E_i / 2kT) \qquad (2.21)$$

where E_i is the energy required to displace an atom from a regular lattice site to an interstitial position [9].

It is known that the presence of defects in a crystalline lattice tends to change the lattice parameter of the crystal. In cubic crystals with a random distribution of defects, the relative lattice parameter change $\frac{\Delta a}{a}$, and the relative change of length of the sample $\frac{\Delta L}{L}$, can be interpreted in terms of the total concentrations of vacant lattice sites C_v, and of occupied interstices, C_i [8,34], according to:

$$3\left(\frac{\Delta L}{L} - \frac{\Delta a}{a}\right) = C_v - C_i \qquad (2.22)$$

The vacancies and not the interstitials that are the dominant point defects in thermal equilibrium [8], so that C_i can be neglected in the above equation. Thus, the change in lattice parameters for pure elements as a function of temperature can be found by:

$$\Delta a(T) = \left(\alpha \Delta T - \frac{C_v(T)}{3}\right) a_o \qquad (2.23)$$

where α is a thermal expansion coefficient, and a_o is the original lattice parameter.

2-6 Lattice parameters versus concentration and temperature

The change in lattice parameters of solid solution due to change in solvent/solute concentrations can be combined with the change due to thermal effect to yield an expression of the lattice parameters of solid solution versus concentration and temperature as:

$$a(X,T) = a_s(X) + \Delta a_A(T).X_A + \Delta a_B(T).X_B \qquad (2.24)$$

where $a_s(X)$ represent the lattice parameters as a function of concentration X, and subscript A and B refer to solvent and solute respectively.

2-7 Computer programming

In order to achieve the calculations proposed in previous sections, a computer programme specifically established for this purpose, was written in Fortran 77.

Functions intended in the programme are:

i- From the atomic size ratio of the elements, calculations may be performed as either substitutional solid solution, interstitial solid solution, or don't form solid solution, as shown in a schematic diagram in Fig. (2-6).

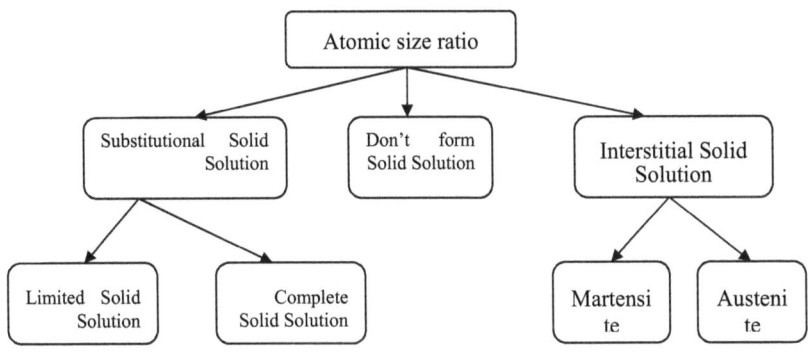

Fig. (2-6) Schematic diagram of the paths followed in the calculations of lattice parameter versus concentration

ii- Conversion from weight percent to atomic percent and/or to number of interstial atoms per 100 solvent atoms is done automatically via equations (2.2), (2.3) and (2.4).

iii- In the case of substitutional solid solution, equations (2.1) and (2.12) are used to obtain the lattice parameters versus concentration.

iv- In the case of interstitial solid solution, if the system is austenite, equation (2.14) is used, and if the system is martensite, equations (2.17) and (2.18) are used to obtain the lattice parameters versus concentration.

v- The change in lattice parameter for pure element due to energy required to form a vacancy and thermal expansion are obtained using equations (2-20) and (2-23) respectively. Finally, the values of lattice parameter of solid solution versus concentration and temperature is obtained using equation (2-24).

Programme flowchart is shown in Fig. (2-7), and programme list is shown in appendix (A).

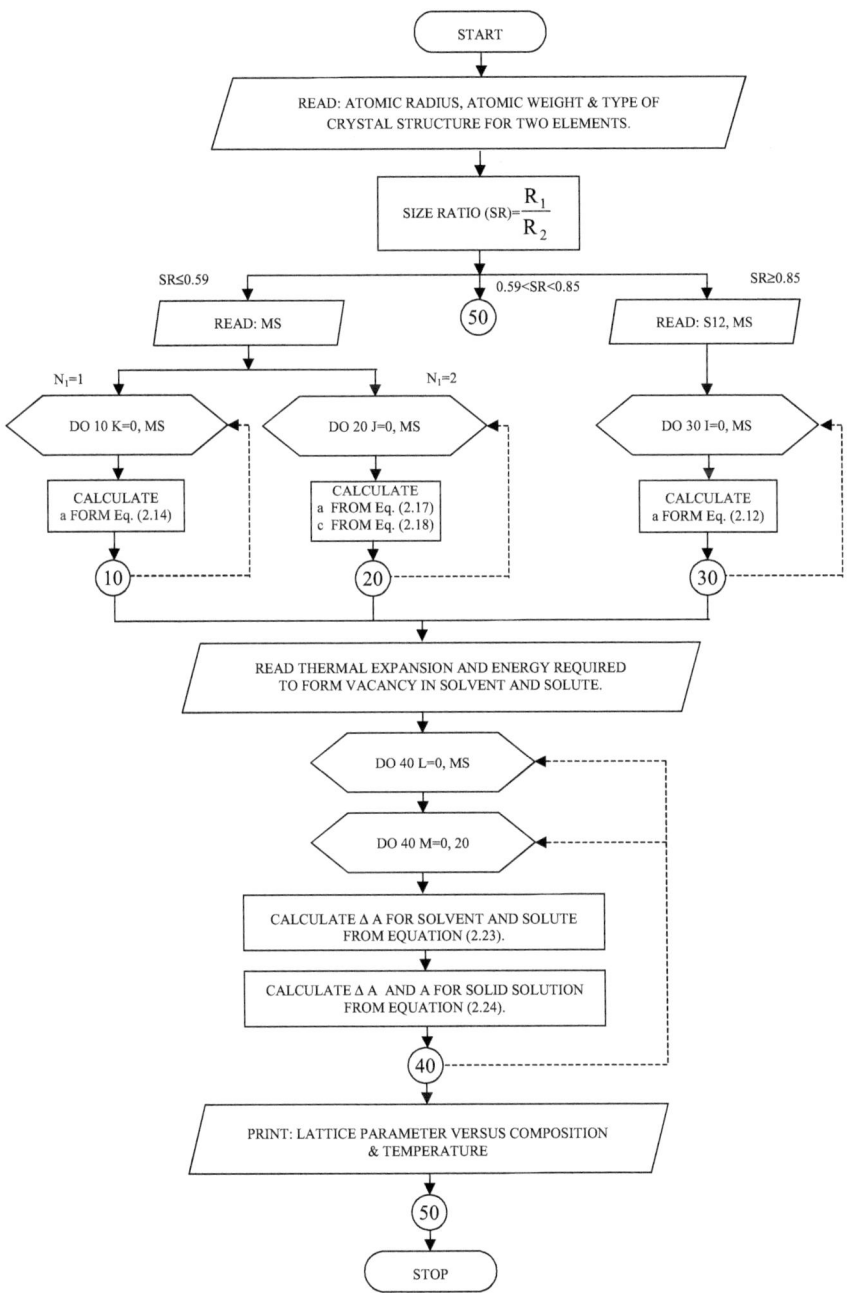

Fig. (2-7) Flowchart of Programme *LAT PAR*.

CHAPTER THREE

Results and Discussion

3-1 Lattice Parameters versus Concentration

3-1-1 The Ni-Cu system

The atomic radii of nickel and copper collected from several references are presented in Table (3-1). The difference of 3.1% in the atomic radius between nickel and copper both face centered cubic structure confirm that formation of substitutional solid solution in most favorable. A case in which complete solubility is to be expected, see Ni-Cu phase diagram (Appendix B).

Table (3-1) Atomic radii of elements of interest extracted from various references.

Atomic radii (nm)					Ref.
Ni	Cu	Si	Ge	C(graphite)	
0.1245	0.1278	0.1176	0.1224	0.071	35
0.12455	0.1278	0.11755	-	0.071	28
0.124595	0.1278	-	0.122485	-	37
0.12435	0.12755	0.11735	0.12225	0.071	3
0.1246	0.1278	-	-	-	10
0.1246	0.127805	0.117585	0.12249	0.07105	5
0.124	0.128	0.117	0.122	-	36
0.124	0.128	0.117	0.122	0.071	1
0.12455	0.1278	0.11755	-	0.071	7

The extent of various atomic radii of nickel and copper on lattice parameters was evaluated by plotting the lattice parameters for the highest and lowest values versus copper concentration as shown in Fig (3.1). The symmetric halves resulted from a line joining a_{Cu} to a_{Ni} (see Fig (3-1)) indicate that average radii in Table (3-1) are the most suitable.

Taking the average radii from all the values in table (3-1) except two reference values (1,36), thus for nickel as r_{Ni} = 0.12453 nm and copper as r_{Cu} = 0.12776 nm, the lattice parameters of Ni-Cu solid solution versus copper concentration were calculated via equation (2.12) as shown in Table (3-2), and are plotted as shown in Fig (3-2), together with the experimental data sited in Moreen with there predicted values [18].

It is evident that the agreement between the prediction model of this work and the experimental data is very good. Moreover, this model yielded lattice parameters that are slightly in better than Moreen's model which may be due to the more realistic values of atomic radii considered in this model.

Table (3-2) Lattice parameters of Ni-Cu solid solution as a function of copper concentration at room temperature.

At. % Cu	Wt. % Cu	Lattice parameters (nm)	
		Vegard's Law	Present work
0	0	0.35224	0.35224
10	10.734	0.35315	0.35302
20	21.295	0.35406	0.32384
30	31.686	0.35498	0.35468
40	41.911	0.35589	0.35555
50	51.975	0.35681	0.35645
60	61.881	0.35772	0.35738
70	71.633	0.35863	0.35834
80	81.235	0.35955	0.35932
90	90.689	0.36046	0.36033
100	100	0.36137	0.36137

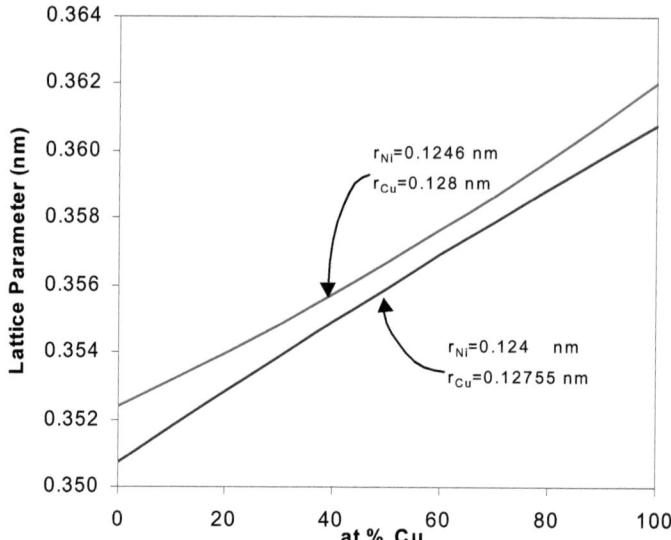

Fig. (3 - 1) Influence of various atomic radii on the lattice parameter of Ni-Cu solid solution.

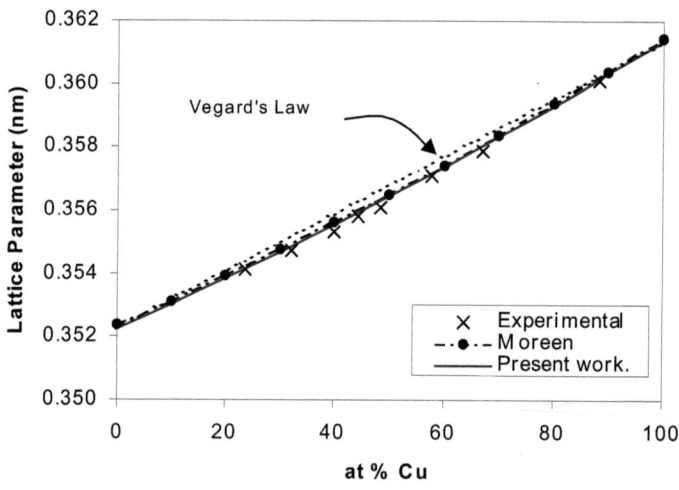

Fig. (3 - 2) Lattice parameter of Ni-Cu solid solution as a function of at% Cu at room temperature.

3-1-2 The Ge-Si system

As Ge-Si solid solution was not included in Moreen's prediction of lattice parameters, it was therefore worth while to evaluate the lattice parameters of Ge-Si system as no such data are found in literature.

From the table (3-1), the difference of 3.7% in the atomic radius between germanium and silicon both diamond structure confirm that the system formed substituational solid solution (see Appendix B).

The influence of various atomic radii on lattice parameters of Ge-Si solid solution is illustrated in Fig (3-3). Lattice parameters of Ge-Si solid solution were calculated using the average radii of germanium and silicon from all the values in table (3-1) except two reference values (1,36), thus (r_{Ge} = 0.122406 nm, r_{Si} = 0.117525 nm). The results are tabulated in Table (3-3). On comparing theoretical values with the experimental [38,39] as shown in Fig (3-4), it is evident that the agreement is very good.

Table (3-3) Lattice parameters of Ge-Si solid solution as a function of silicon concentration at room temperature.

At. % Si	Wt. % Si	Lattice parameters (nm)	
		Vegard's Law	Present work
0	0	0.56537	0.56537
10	4.121	0.56312	0.56294
20	8.819	0.56086	0.56054
30	14.223	0.55861	0.55819
40	20.505	0.55635	0.55587
50	27.897	0.55410	0.55360
60	36.727	0.55185	0.55137
70	47.445	0.54959	0.54917
80	60.748	0.54734	0.54702
90	77.689	0.54509	0.54491
100	100	0.54283	0.54283

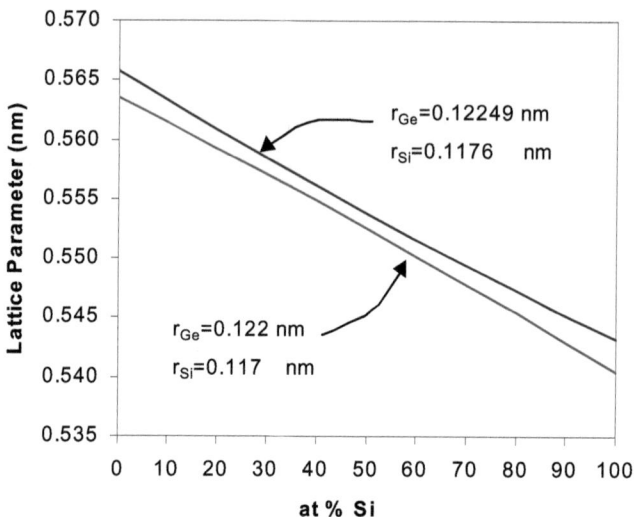

Fig. (3-3) Influence of various atomic radii on the lattice parameter of Ge-Si solid solution.

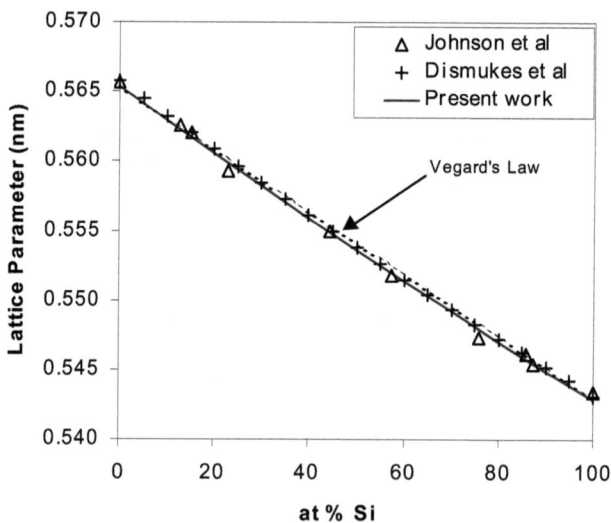

Fig. (3-4) Lattice parameters of Ge-Si solid solution as a function of at% Si at room temperature.

3-1-3 The Fe-C system

The atomic radii for the phases of iron collected from several references are presented in Table (3-4).

Table (3-4) Atomic diameters (*d*) and atomic radii(*r*) of iron for all allotropic forms, extracted from various references.

Iron phase	*d*(nm)	*r*(nm)	Ref.
α	0.2481	0.12405	
γ 908<T<1405°C	0.2585	0.12925	28
δ T>1403 °C	0.254	0.127	
α T=20°C	0.24823	0.124115	
γ T=916°C	0.25786	0.12893	37
δ T=1394°C	0.25393	0.126965	
α	0.2477	0.12385	
γ T=916°C	0.2575	0.12865	3
δ T=1387°C	0.2534	0.1267	
α T=20°C	0.24824	0.12412	
γ T=916°C	0.25787	0.128935	5
δ T=1394°C	0.25394	0.12697	
bcc	0.2482	0.1241	10
fcc	0.2522	0.1261	
fcc	0.252	0.126	36
α(bcc)	0.24824	0.12412	7

The ratio of atomic radius of graphite to that of iron is less than 0.59 an indication that the size ratio is favorable to the formation of interstitial solid solution, i.e. carbon atoms enter the holes that exist in the iron structure as:

i- Austenite:

In this case iron atoms have face centered cubic structure, therefore the atomic radius of γ-Fe (0.1261 nm) and the atomic radius of graphite (0.071 nm) were employed in the calculations. According to equation (2.14) the values of lattice parameter for Fe-C austenite as a function of carbon concentration shown in Table (3-5) are plotted together with the experimental data as illustrated in Fig (3-5). It is clear that linear relationship exist between lattice parameters and carbon concentration.

Table (3-5) Lattice parameters of Fe-C austenite as a function of carbon concentration at room temperature, according to equation (2.14).

C/100 Fe	C at. %	C wt. %	a(nm)
0	0	0	0.35666
1	0.990	0.214	0.35760
2	1.960	0.428	0.35854
3	2.912	0.641	0.35948
4	3.846	0.853	0.36042
5	4.762	1.064	0.36136
6	5.660	1.274	0.36230
7	6.542	1.483	0.36323
8	7.407	1.691	0.36417
9	8.256	1.898	0.36511
10	9.091	2.105	0.36605

Comparing the theoretical results of this work with the relevant experimental results [24,40], it is found that the theoretical values of lattice parameters are generally larger than the experimental values over the entire carbon concentration 10 C/100 Fe, (see Fig (3-5)). This is due to the assumption made that the specimen is pure austenite and has regular structure; whereas in practice, formed austenite at room temperature is usually prepared by very rapid quenching in brine or liquid nitrogen. This sort of preparation procedure undoubtedly causes crystal defects. The concentrations of these defects depend upon the starting and final temperature of quenching and time taken. Thus the increase of defects concentration, whether vacancy or interstitial, causes the apparent decrease in the lattice parameters.

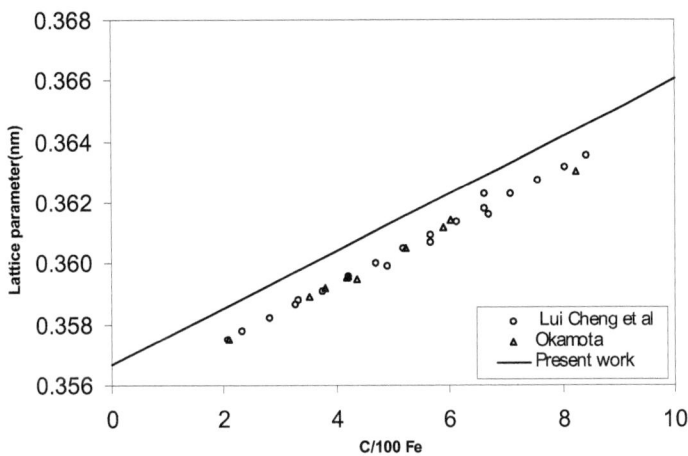

Fig.(3 - 5) Lattice parameters of Fe-C austenite as a function of carbon concentration at room teperature.

ii- Martensite:

In this case the atomic radius of α-Fe (0.1241 nm), and the atomic radius of graphite (0.071 nm) were used. According to equations (2.17) and (2.18), the values of the lattice parameters **a** and **c** for Fe-C martensite as a function of carbon concentration shown in Table (3-6) are plotted together with the experimental data as illustrated in Fig (3-6). It is clear that linear relationship exist between *a* and *c*, and C/100 Fe, but with *a* decreasing and *c* increasing by different degrees.

Table (3-6) Lattice parameters of Fe-C martensite as a function of carbon concentration at room temperature.

C/100 Fe	C at. %	C wt. %	a(nm)	c(nm)
0	0	0	0.28660	0.28660
1	0.990	0.214	0.28633	0.28919
2	1.960	0.428	0.28606	0.29178
3	2.912	0.641	0.28579	0.29437
4	3.846	0.853	0.28553	0.29696
5	4.762	1.064	0.28526	0.29955
6	5.660	1.274	0.28499	0.30214
7	6.542	1.483	0.28473	0.30473
8	7.407	1.691	0.28446	0.30732
9	8.256	1.898	0.28419	0.30991
10	9.091	2.105	0.28392	0.31250

Comparing the theoretical results with the experimental work by Cadeville et al. [32], Bernshtein et al. [41], and experimental results gathered by Liu Cheng et al. [24], and Okamoto [40], it is found that the theoretical values of a-axis parameter are in a very good agreement with the experimental values, whereas the values of c-axis parameter are slightly higher than the experimental results (see Fig (3-6)).

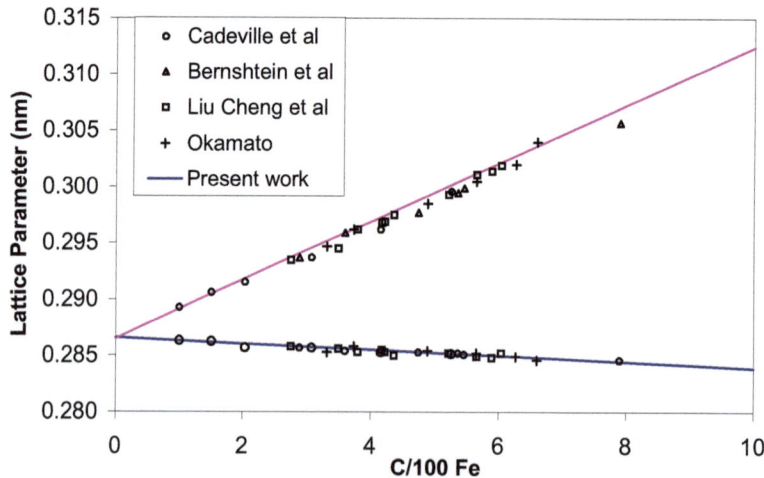

Fig.(3-6) Lattice parameters of Fe-C martensite as a function of carbon concentration at room temperature.

Explanation may be proceeded as follows:

In practice, the martensite is prepared by means of quenching in brine water followed by processes of heat treatment, which would then cause reduction in crystal defects and relief of stresses between the atoms; thereby no significant deviation is observed between experimental and theoretical values. Aging also play a role in the redistribution of carbon atoms [42]; and that aged specimens lead to decrease of the c parameter, and minute decrease in a parameter [43-45].

On the other hand, it must not be forgotten that the experimental data have also a large scatter from each other. The scatter can have several origins; for example, impurities dissolved in the martensite matrix, experimental errors due to different methods used for lattice parameter determination, and temperature uncertainty that pertain to lattice parameter data [24].

3-2 Lattice parameter versus temperature

Materials properties are much influenced by temperature, which necessitate the knowledge of the changes of lattice parameters of a solid solution with temperature in addition to concentration subject to the conditions that the change in lattice parameters via temperature was assumed to be isotropic, and the effect of interstitial defects are neglected.

3-2-1 The Ni-Cu system

The change in lattice parameters with temperature for each element was calculated separately. Thermal expansion coefficients (α_T) presented in Table (3-7), were used to find the relative change in length ($\frac{\Delta L}{L}$).

Table (3-7) Thermal expansion coefficients of relevant elements.

$$\alpha_T = A + B(T-T_o) + C(T-T_o)^2$$

Element	T_o (K)	A (10^{-6}/K)	B (10^{-9}/K^2)	C (10^{-11}/K^3)	Range of temperatures (K)	Ref.
Nickel	273	12.87	7.64	0.036	320-1370	46,47
Copper	273	16.872	1.773	1.146	323-623	47,48
Iron(γ)	-	24.7	0	0	T$\geq$1000	49
Iron(α)	273	11.45	14.0	-1.089	272-970	47
Graphite		1.3	0	0	273-773	47
		2.2	0	0	773-1273	
		2.6	0	0	1273-1773	
		3.2	0	0	1773-2573	

From the energy required to form a vacancy (1.74 ev for nickel and 1.24 ev for copper) [28], the number of vacancies and vacancy concentrations were also found. The change in lattice parameters (Δa) for nickel and copper as a function of temperature were then calculated according to the equation (2.23) and are presented in Table (3-8).

Table (3-8) Change in lattice parameters of pure nickel and copper with temperature.

T(K)	Nickel $\frac{\Delta L}{L}$	Nickel Δa(nm)	Copper $\frac{\Delta L}{L}$	Copper Δa(nm)
300	0	7.327 E-31	0	1.854 E-22
400	1.384 E-03	4.877 E-04	1.728 E-03	6.245 E-04
500	2.924 E-03	1.030 E-03	3.573 E-03	1.291 E-03
600	4.621 E-03	1.627 E-03	5.603 E-03	2.024 E-03
700	6.478 E-03	2.282 E-03	7.887 E-03	2.850 E-03
800	8.498 E-03	2.993 E-03	1.0494 E-02	3.792 E-03
900	1.0680 E-02	3.762 E-03	1.3493 E-02	4.876 E-03
1000	1.3029 E-02	4.589 E-03	1.6952 E-02	6.126 E-03
1100	1.5547 E-02	5.476 E-03	2.0940 E-02	7.567 E-03
1200	1.8234 E-02	6.423 E-03	2.5525 E-02	9.223 E-03
1300	2.1095 E-02	7.430 E-03	3.0780 E-02	1.1121 E-02

The lattice parameters of Ni-Cu solid solution in terms of concentration and temperature are obtained using equation (2.24). Values which are tabulated in Appendix C1 are plotted, firstly as lattice parameters versus copper concentrations for different temperatures (Fig (3-7)), and secondly as lattice parameters versus temperatures for different copper concentrations (Fig (3-8)). Also three dimensional plot is shown in (Fig.(3-9)).

It can be seen the lattice parameters of Ni-Cu solid solution increases non linearly with the increase of temperature; the increase in the copper side being greater than the increase in the nickel side. This is because the thermal expansion coefficient of copper is greater than that of nickel at the same temperature, moreover the vacancy concentrations, which are responsible for the changes in lattice parameters, are higher in copper in comparison with nickel.

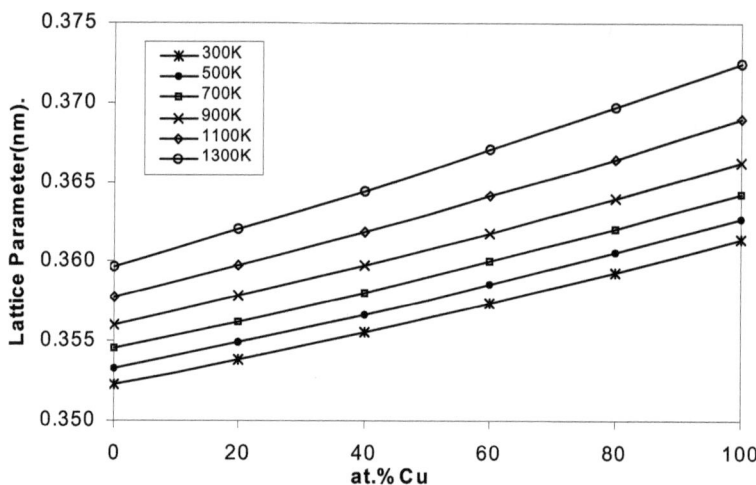

Fig.(3 - 7) Lattice parameters of Ni-Cu solid solution as a function of at % Cu for different temperatures.

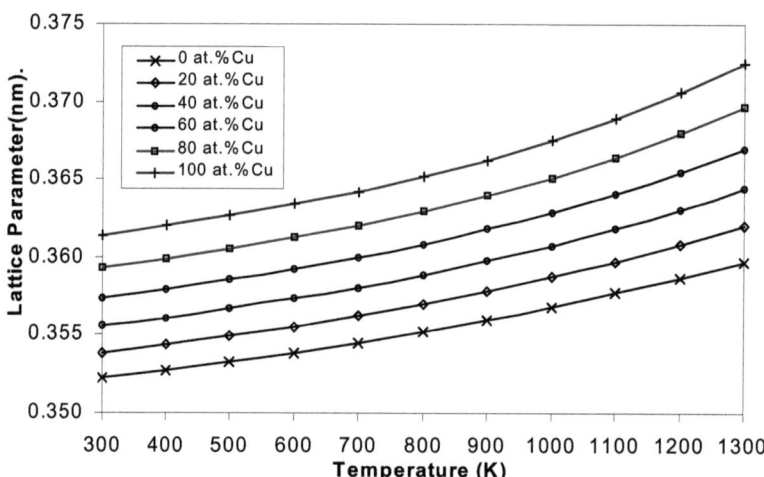

Fig.(3 - 8) Lattice parameters of Ni-Cu solid solution as a function of Temperature, for different concentrations.

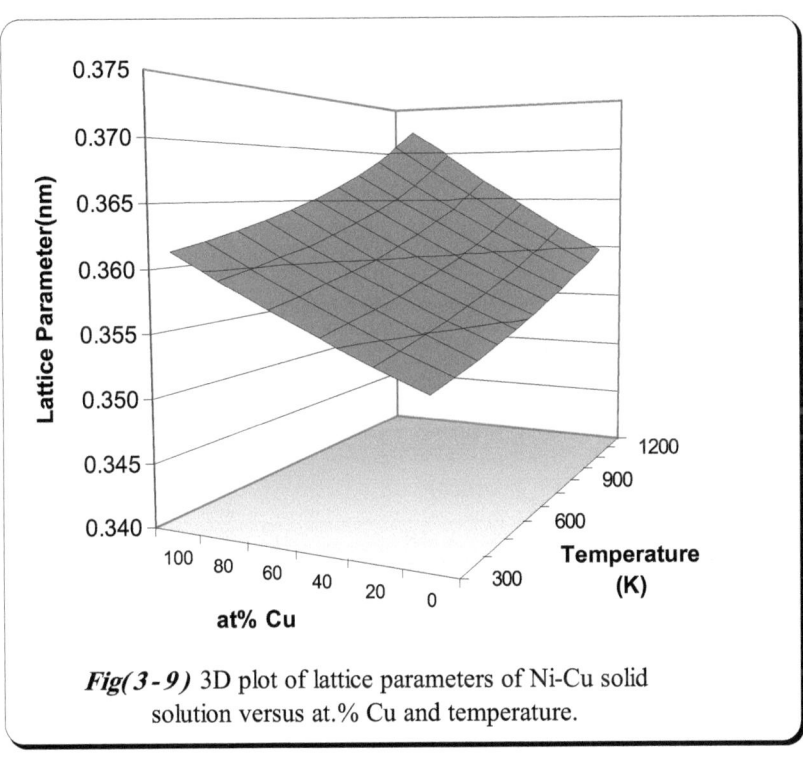

Fig(3-9) 3D plot of lattice parameters of Ni-Cu solid solution versus at.% Cu and temperature.

3-2-2 The Fe-C system

i- Austenite:

For the sake of obtaining changes in lattice parameters for Fe-C austenite via temperature, the same calculations sited in 3-2-1 were followed. The change in lattice parameters for pure γ-iron and graphite as a function of temperature are presented in Table (3-9).

Table (3-9) Change in lattice parameters of graphite and γ–iron with temperature.

T(K)	Carbon(graphite) $\frac{\Delta L}{L}$	Carbon(graphite) Δa(nm)	γ-Iron $\frac{\Delta L}{L}$	γ-Iron Δa(nm)
300	0	1.348 E-18	0	1.953 E-18
400	1.30 E-04	3.199 E-05	2.470 E-03	8.809 E-04
500	2.60 E-04	6.399 E-05	4.940 E-03	1.761 E-04
600	3.90 E-04	9.599 E-05	7.410 E-03	2.642 E-03
700	5.20 E-04	1.279 E-04	9.880 E-03	3.523 E-03
800	1.10 E-03	2.707 E-04	1.235 E-02	4.404 E-03
900	1.32 E-03	3.246 E-04	1.482 E-02	5.285 E-03
1000	1.54 E-03	3.782 E-04	1.729 E-02	6.165 E-03
1100	1.76 E-03	4.310 E-04	1.976 E-02	7.044 E-03
1200	1.98 E-03	4.821 E-04	2.223 E-02	7.921 E-03
1300	2.60 E-03	6.289 E-04	2.470 E-02	8.793 E-03
1400	2.86 E-03	6.831 E-04	2.717 E-02	9.660 E-03
1500	3.12 E-03	7.318 E-04	2.964 E-02	1.0519 E-02
1600	3.38 E-03	7.734 E-04	3.211 E-02	1.1367 E-02
1700	3.64 E-03	8.063 E-04	3.458 E-02	1.2203 E-02

Similarly, the lattice parameter of Fe-C austenite in terms of carbon concentration and temperature calculated via the equation (2.24), yielded the values listed in Appendix C2 which are plotted firstly as lattice parameter with C/100Fe for different temperature (Fig (3-10)), and secondly as lattice parameter with temperature for different carbon concentration (Fig (3-11)). Also three dimensional plot is shown in (Fig.(3-12)).

It can be seen that the lattice parameters of Fe-C austenite increase linearly as the carbon concentration and temperature increases. For the purpose of comparison, the experimental data reported by Onink [25] was also plotted on Fig (3-11). It is thus evident that the agreement between the theoretical of this study and the experimental data is excellent.

However the difference observed between theoretical and experimental values of lattice parameters versus concentration of austenite at room temperature (see Fig(3-5)), which was interpreted due to defects and stresses that resulted from preparation condition, implies that the model of prediction is correct and the apparent difference was due to departure from regular structure of the pure austenite.

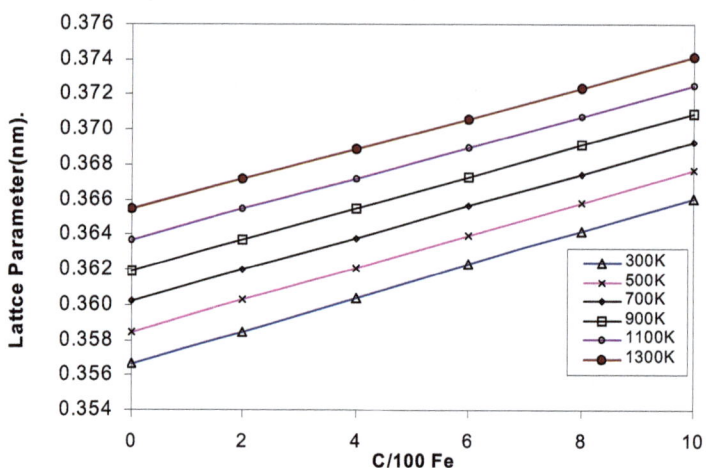

Fig.(3 - 10) Lattice Parameters of austenite as a function of the carbon concentration for different temperatures.

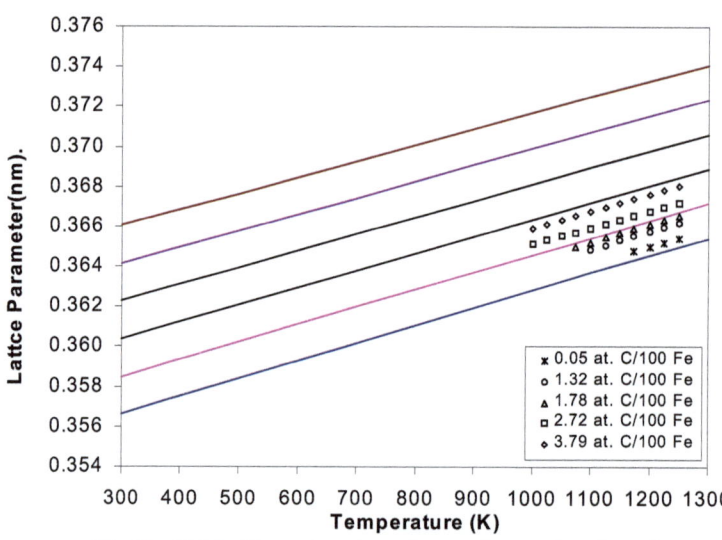

Fig.(3 - 11) Lattice parameters of austenite as a function of temperature for different concentrations.

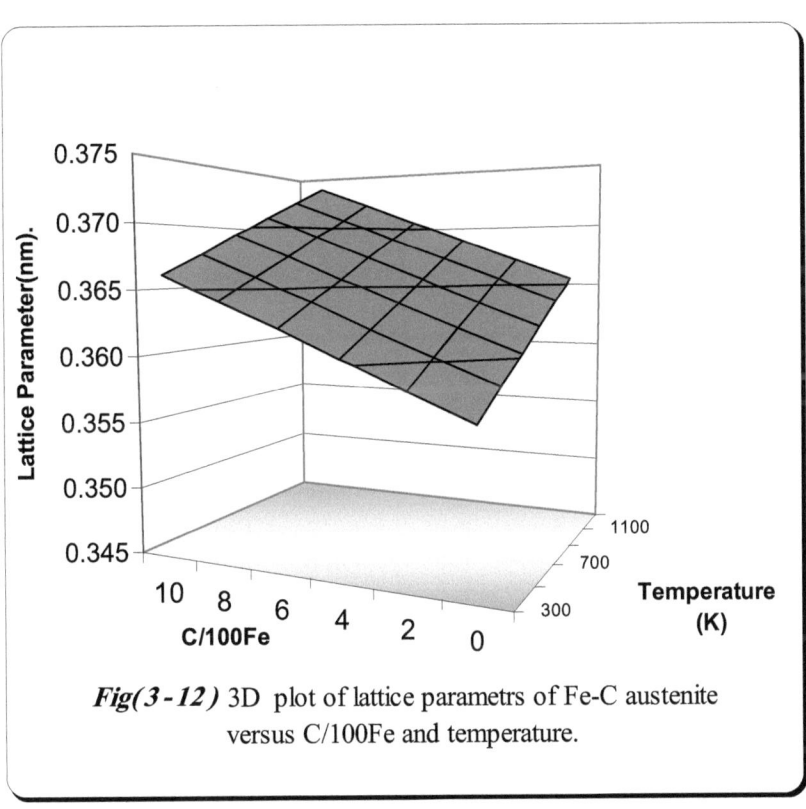

Fig(3-12) 3D plot of lattice parametrs of Fe-C austenite versus C/100Fe and temperature.

ii- Martensite:

The change in a and c parameters of the body centered tetragonal structure were done as in the case of austenite described previously, except using thermal expansion coefficient for α-Fe instead of γ-Fe. The change in lattice parameters for pure α-iron and graphite as a function of temperature, are presented in Table (3-10).

Table (3-10) Change in lattice parameters of graphite and α–iron with temperature.

T(K)	Carbon(graphite)		α-Iron	
	$\frac{\Delta L}{L}$	Δa(nm)	$\frac{\Delta L}{L}$	Δa(nm)
300	0	1.348 E-18	0	1.569 E-18
400	1.30 E-04	3.199 E-05	1.305 E-03	3.740 E-04
500	2.60 E-04	6.399 E-05	2.813 E-03	8.062 E-04
600	3.90 E-04	9.599 E-05	4.458 E-03	1.277 E-03
700	5.20 E-04	1.279 E-04	6.176 E-03	1.770 E-03
800	1.10 E-03	2.707 E-04	7.901 E-03	2.264 E-03
900	1.32 E-03	3.246 E-04	9.567 E-03	2.741 E-03
1000	1.54 E-03	3.782 E-04	1.1114 E-02	3.184 E-03
1100	1.76 E-03	4.310 E-04	1.2464 E-02	3.569 E-03

Similarly, the lattice parameters (a and c) of the Fe-C martensite in terms of carbon concentration and temperature are shown in Appendices C3 and C4. Figs.(3-13), and (3-14) illustrate the relation between lattice parameters, via C/100 Fe for different temperatures, and via temperature for different concentrations, respectively. And three dimensional plot is shown in Figs.(3-15) and (3-16). In this case, it can bee seen that the a and c axes of the unit cell increases linearly as the carbon concentration and temperature increases.

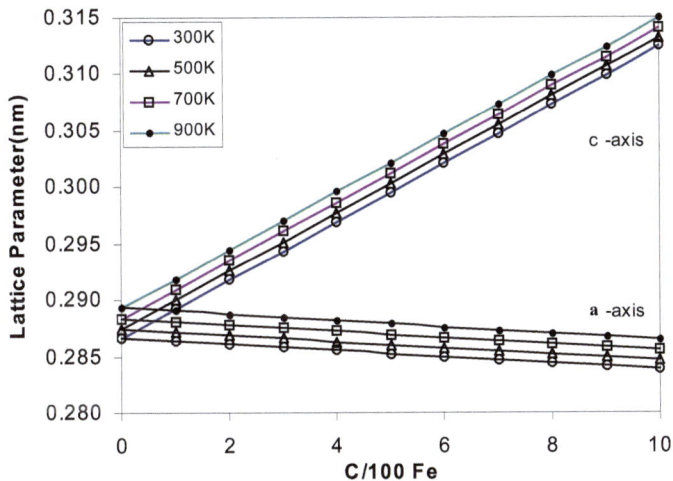

Fig.(3 - 13) Lattice parameters of martensite as a function of the carbon concentration for different temperatures.

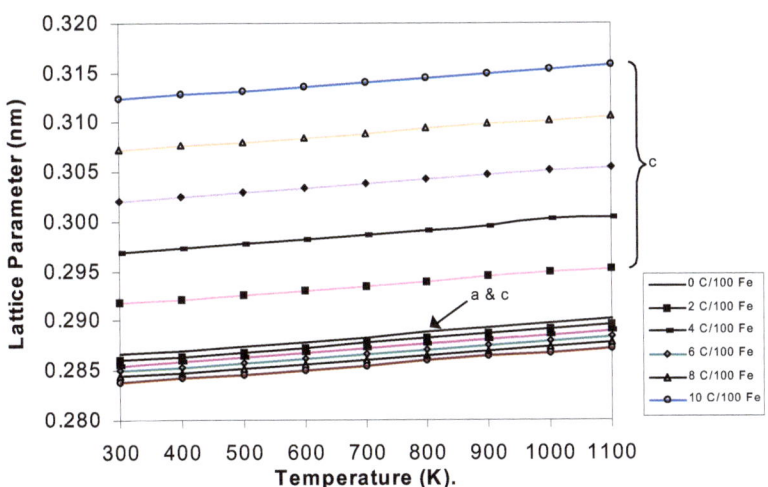

Fig. (3 - 14) Lattice parameters of martensite as a function of temperature for different concentrations.

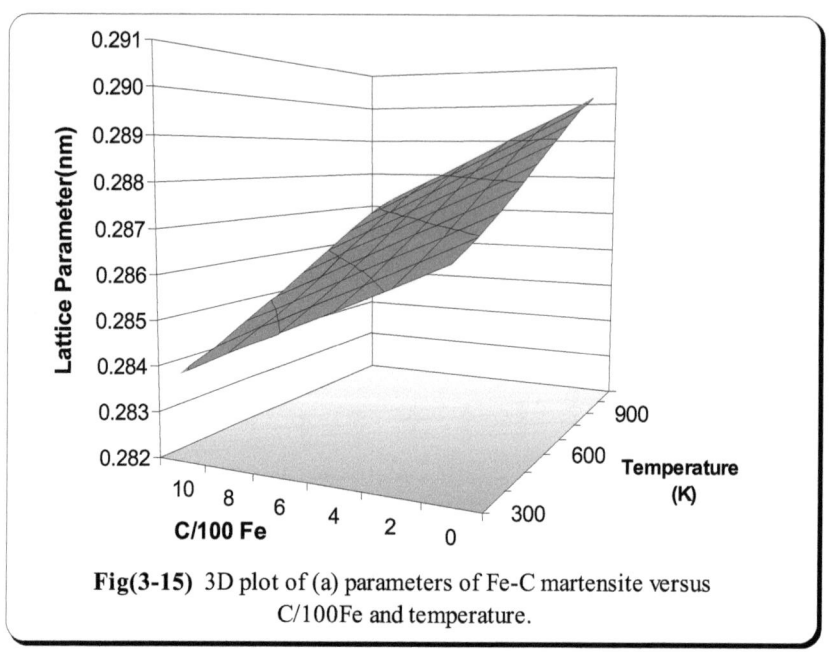

Fig(3-15) 3D plot of (a) parameters of Fe-C martensite versus C/100Fe and temperature.

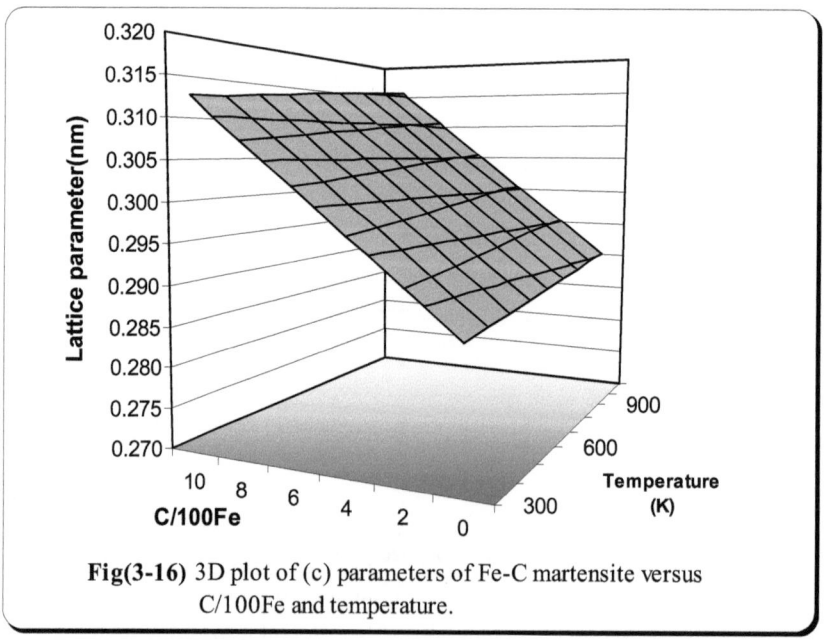

Fig(3-16) 3D plot of (c) parameters of Fe-C martensite versus C/100Fe and temperature.

iii- Ferrite:

When the carbon concentration in α-Fe is zero, the case is represented by the ferrite solid solution; the lattice parameters of ferrite is independent of carbon concentration, because of much more restricted solubility of carbon (maximum of 0.025 wt% C) [50,51].

Comparing the lattice parameters in this case as a function of temperature only, it is evident that the agreement between the theoretical results of this study and the experimental data [25] is very good (see Fig(3-17)).

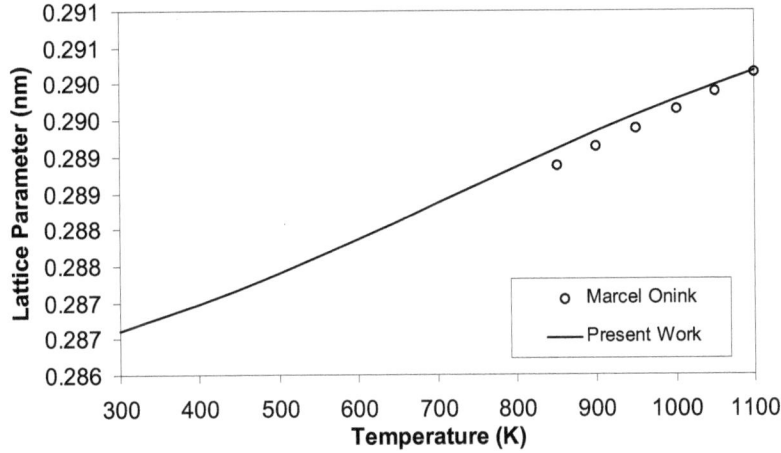

Fig(3-17) Lattice parameters of ferrite as a function of temperature.

CHAPTER FOUR

Conclusions and Suggestions for future work

4-1 Conclusions

i- Substituational solid solutions:

The model described by Moreen [18] has been extended by introducing atomic radii of the binary components rather than the interatomic spacing to predict lattice parameters of solid solution alloys. This model had the advantage of the possibility of the proper selection of the atomic radii that can optimize the agreement with the experimental data. The combination of the temperature effect via vacancy defects and thermal expansion with the concentration has given the model a new general from.

ii- Interstitial solid solutions:

The models described by equation (2.14) for austenite Fe-C and equations (2.17) and (2.18) for martensite Fe-C have been proposed to calculate the lattice parameters of the interstitial solid solution alloys in terms of carbon content. The models are also extended to account for thermal effect via equation (2.24). A comparison in terms of carbon content of data from several sources for martensite Fe-C has shown very good agreement with the calculated data. Some deviations from the predicted parameter values are noted in the room temperature austenite Fe-C where very rapid quenching procedures are known to cause crystal defects. Like wise a comparison in terms of temperature of data from available sources for austenite and ferrite has also shown very good agreement.

4-2 Suggestions for future work

The following points can be taken as a guide lines for future work:

i- An attempt is worth while to extend the equation ($a^{(ave)} = \sum_{i=1}^{n}\sum_{j=1}^{n} a_{ij}\, c_i c_j$) to calculate lattice parameters of the ternary (Cu-Ni-Mn) or (Cu-Ni-Cr) system.

ii- Combination of the equations in substituational and interstitial solid solutions, to predict the lattice parameters of ternary system such as (Mn-Fe-C) is very useful.

iii- Concerning the term S_{12} stated in chapter two, it is of importance to try and find an approach to predict the separation of unlike atoms on theoretical grounds.

REFERENCES

[1] Leonid V. Azaroff, "Introduction to Solids", McGraw-Hill Inc., New York (1997).

[2] J.S. Blakemore, "Solid State Physics", Second Edition, W.B.Saunders Company (1974).

[3] William Hume-Rothery, R.E. Smallman, and C.W.Haworth, "The Structure of Metals and Alloys", The Institute of Metals (1969).

[4] A.F. Wells, "Structural Inorganic Chemistry", Third Edition, Oxford University Press (1962).

[5] B.D. Cullity, "Elements of X-Ray Diffraction", Second Edition, Addison-Wesley Publishing Company, Inc. (1978).

[6] G.K. Narula, K.S. Narula, and V.K. Gupta, "Materials Science", Tata McGraw-Hill Publishing Company Limited (1997).

[7] Sidney H. Avner, "Introduction to Physical Metallurgy", Second Edition, Tata McGraw-Hill Publishing Company Limited (1997).

[8] A. Seeger, D. Schumacher, W. Schilling, and J. Diehl, "Vacancies and Interstitials in Metals", North-Holland Publishing Company, Amsterdam (1970).

[9] Charles Kittel, "Introduction to Solid State Physics", Sixth Edition, John Wiley & Sons. Inc., New York (1986).

[10] W.B. Pearson, "The Crystal Chemistry and Physics of Metals and Alloys", John Wiley & Sons, Inc. (1972).

[11] Y.C. Venudher, Leela Iyengar, and K.V. Krishna Rao, "Thermal Expansion and Debye Temperatures of KCl-KBr Mixed Crystals to an X-Ray Method", Journal of Materials Science 21, 110-116 (1986).

[12] G.P. Kostikova, and Yu.P. Kostikov, "Application of Vegard's Law to the Description of Solid Solutions", Inorganic Materials 29, 1136-1137 (1993).

[13] M.A. Jaswon, W.G. Henry, and G.V. Raynor, "The Cohesion of Alloys: I.Intermetallic Systems Formed by Copper, Silver and Gold, and Deviations from Vegard's Low", Proc. Phys. Soc., Lond. B, 64, 177-189 (1951). (Phys. Abst.51-3677).

[14] E-An Zen, "Validity of Vegard's Law", Am. Mineralogist. 41, 523-524 (1956).

[15] G.J. Dienes, "Lattice Parameter and Short-Range Order", Acta Metallurgica 6, 278-282 (1958).

[16] F.M. d'Heurle, A.S. Nowick, and D.P. Seraphim, Acta Crystal. 13, 1009 (1960).

[17] K.A. Gschneidner, Jr., and G.H. Vineyard, "Departures from Vegard's Law", Journal of Applied Physics 33, 3444-3450 (1962).

[18] H.A. Moreen, R. Taggart, and D.H. Polonis, "A Model for the Prediction of Lattice Parameters of Solid Solutions", Metallurgical Transactions 2, 265-268 (1971).

[19] Ning Yuantao, Xu. Hua, "An Approach to Predicting Lattice Parameters of Certain Solid Solution", Acta Metall. Sin. 21, B45-50 (1985).

[20] M. Ron, A. Kidron, H. Schechter, and S. Niedzwiedz, "Structure of Martensite", Journal of Applied Physics 38, 590-594 (1967).

[21] S.C. Moss, "Static Atomic Displacements in Iron-Carbon Martensite", Acta Metallurgica 15, 1815-1826 (1967).

[22] G.V. Kurdjumov and A.G. Khachaturyan, "Nature of Axial Ratio Anomalies of the Martensite Lattice and Mechanism of Diffusionless $\gamma \rightarrow \alpha$ Transformation", Acta Metallurgica 23, 1077-1088 (1975).

[23] M. Takahashi, K. Nushiro, and S. Ishio, "Formation of The F.C.C. Phase in Fe-C Alloys by Rapid Quenching", Phys. Stat. Sol. (a) 89, K27-29 (1985).

[24] Liu Cheng, A. Böttger, Th. H. de Keijser, and E.J. Mittemeijer, "Lattice Parameters of Iron-Carbon and Iron-Nitrogen Martensites and Austenites", Scripta Metallurgica et Materialia 24, 509-514 (1990).

[25] M. Onink, "Decomposition of Hypo-eutectoid Iron Carbon Austenites", Ph.D. Thesis, Delft University of Technology (1995).

[26] S.O. Pillai, "Solid State Physics", New Age International (P) Limited Publishers (1997).

[27] Harald Ibach and Hans Lüth, "Solid State Physics", Narosa Publishing House (1996).

[28] Ropert M. Brick, Robert B. Gordon, and Arthur Phillips, "Structure and properties of alloys", Third Edition, McGraw-Hill, Inc. (1966).

[29] R.E. Smallman, "Modern Physical Metallurgy", Fourth Edition, Butterworth &Co (Publishers) Ltd. (1985).

[30] D.H. Jack, and K.H. Jack, "Invited Review: Carbides and Nitrides in Steel", Materials Science and Engineering 11, 1-27 (1973).

[31] C.P. Flynn, "Point Defects and Diffusion", Oxford University Press (1972).

[32] M.C. Cadeville, J.M. Friedt, and C. Lerner, "Structural, Electronic and Magnetic Properties of Splat-Quenched FeC_x Alloys ($x \leq 0.05$)" J. Phys. F: Metal Phys. 7, 123-137 (1977).

[33] W.K. Choo, and Ray Kaplow, "Mössbauer Measurements on the Aging of Iron-Carbon Martensite", Acta Metallurgica 21, 725-732 (1973).

[34] B. Henderson, "Defects In Crystalline Solids", Edward Arnold (Publishers) Ltd. (1972).

[35] Lawrence H. Van Vlack, "A textbook of materials technology", Addison-Wesley Publishing Company, Inc. Philippines Copyright (1973).

[36] J.G. Stark, and H.G. Wallace, "Chemistry Data Book", Second Edition in SI, J.G. Stark, H.G. Wallace (1982).

[37] W.B. Pearson, "A Handbook of Lattice Spacings and Structures of Metals and Alloys", Volume 2, Pergamon Press Ltd., Headington Hill Hall, Oxford (1967).

[38] Everett R. Johnson, and Schuyler M. Christian, "Some Properties of Germanium-Silicon Alloys", Physical Review 95, 560-561 (1954).

[39] J.P. Dismukes, L. Ekstrom, and R.J. Paff, "Lattice Parameter and Density in Germanium-Silicon Alloys", The Journal of Physical Chemistry 68, 3021-3027 (1964).

[40] H. Okamoto, "The C-Fe (Carbon-Iron) System", Journal of Phase Equilibrium 13, 543-565 (1992).

[41] M. L. Bernshteyn, L. M. Kaputkina, and S.D. Prokoshkin, "Martensite structure after the $\kappa' \rightarrow \alpha$ transition", Phys. Met. Metall. 52, 127-139 (1981).

[42] Jean-Marie R. Genin, "The Clustering and Coarsening of Carbon Multiples during the Aging of Martensite from Mössbauer Spectrosco The Precipitation Stage of Epsilon Carbide", Metallurgical Transactions 18A, 1371-1388 (1987).

[43] Liu Cheng, C.M. Brakman, B.M. Korevaar, and E.J. Mettemeijer, "The Tempering of Iron-Carbon Martensite; Dilatometric and Calorimetric Analysis", Metallurgical Transactions 19A, 2415-2426 (1988).

[44] Lui Cheng, N.M. van der Pers, A. Böttger, Th.H.de Keijser, and E.J. Mittemeijer, "Lattice Changes of Iron-Carbon Martensite on Aging at Room Temperature", Metallurgical Transactions 22A, 1957-1967 (1991).

[45] A. Böttger, and E.J. Mittemeijer, "Redistribution of interstitial atoms on aging of iron-based martensites", Speich Symposium Proceedings, 5-17 (1992).

[46] T.G. Kollie, "Measurement of the thermal-expansion coefficient of nickel from 300 to 1000 K and determination of the power law constant near the curie temperature", Physical Review B16, 4872-4881 (1977).

[47] R.S. Krishnan, R. Srinivasan, and S. Devanarayanan, "Thermal Expansion of Crystals", First Edition, Pergamon Press Ltd. (1979).

[48] S.J. Bennett, "The Thermal Expansion of Copper Between 300 and 700 K", J. Phys. D: Appl. Phys. 11, 777-780 (1978).

[49] M. Onink, C.M. Brakman, F.D. Tichelaar, E.J. Mittemeijer, S. van der Zwaag, J.H. Root, and N.B. Konyer, "The Lattice Parameters of Austenite and Ferrite in Fe-C Alloys as Functions of Carbon Concentration and Temperature", Scripta Metallurgica Et Materialia 29, 1011-1016 (1993).

[50] S.G. E. te Velthhuis, J.H. Root, J.Sietsma, M.Th. Rekveld, and S. van der Zwaag, "The Ferrite and Austenite Lattice Parameters of Fe-Co and Fe-Cu Binary alloys as a function of temperature", Acta Mater. 46, 5223-5228 (1998).

[51] N.N. Rammo, and O.G. Abdullah, "A model for the prediction of lattice parameters of iron–carbon austenite and martensite", Journal of Alloys and Compounds, 420, 117-120, (2006).

APPENDIX A

Listing of the programme LAT PAR

```
C ---- LATTICE PARAMETERS AS A FUNCTION OF COMPOSITION ----------------
      DIMENSION SS(0:200),AA(0:200,0:200),CC(0:200,0:200)
      DIMENSION DA1(0:100),DA2(0:100)
      OPEN (5,FILE='IA')
      OPEN (6,FILE='OA')
      READ (5,*)R1,R2
      IF(R1.LE.R2)GOTO 20
      SR=R2/R1
      GOTO 30
  20  SR=R1/R2
  30  READ(5,*)W1,W2
      READ(5,*)N1,N2
      IF(SR.GE.0.85)GOTO 40
      IF (SR.LT.0.59)GOTO 200
      GOTO 500
C ---- SUBSTITUATIONAL SOLID SOLUTION --------------------------------
  40  MM=5
      CALL LATTICE(R1,N1,A1,S11)
      CALL LATTICE(R2,N2,A2,S22)
      READ(5,*)s12
      IF(N1.EQ.N2)GOTO 70
      READ (5,*)MS
      GOTO 80
  70  MS=100
  80  DO 90 I=0,MS,MM
      C=FLOAT (I)/100.
      SS(I)=(1.-C)**2*S11+2.*(1.-C)*C*S12+C**2*S22
  90  CONTINUE
      CALL ALLOY (SS,AA,MS,N1)
      WRITE(6,100)
 100  FORMAT (4X,'ATOMIC%',5X,'WEIGHT%',5X,'LATT.PARA.')
      WRITE (6,110)
 110  FORMAT (4X,'-------',5X,'-------',5X,'----------')
      DO 130 N=0,MS,MM
      WP=(100.*N*W2/W1)/(100.+(W2/W1-1.)*N)
      WRITE(6,120)N,WP,AA(N,0)
 120  FORMAT(4X,I5,5X,F8.3,5X,F8.3)
 130  CONTINUE
      GOTO 400
C ---- INTERSTITIAL SOLID SOLUTION ------------------------------------
 200  MM=1
      READ(5,*)MS
      GOTO(210,270),N1
 210  DO 220 I=0,MS
```

```
      F=2.*SQRT(2.)*R1
      AA(I,0)=F+((2.*(R1+R2)-F)/(4.*FLOAT(MS))) *FLOAT(I)
  220 CONTINUE
      WRITE(6,230)
  230 FORMAT(4X,'ATOMIC%',5X,'WEIGHT%',5X,'SV/100 SV',5X,'LATT.PARA.')
      WRITE(6,240)
  240 FORMAT(4X,'-------',5X,'-------',5X,'---------',5X,'----------')
      DO 260 N=0,MS
      WP=(100.*N*W2/W1)/(100.+(W2/W1-1.)*N)
      AP=(100.*N)/(100-N)
      WRITE(6,250)N,WP,AP,AA(N,0)
  250 FORMAT(4X,I5,5X,F8.3,5X,F8.3,5X,F8.3)
  260 CONTINUE
      GOTO 400
  270 DO 280 J=0,MS
      F=4.*R1/SQRT(3.)
      AA(J,0)=F-((F-SQRT(2.)*(R1+R2))/(4.*FLOAT(MS)))*FLOAT(J)
      CC(J,0)=F+((2.*(R1+R2)-F)/(4.*FLOAT(MS)))*FLOAT(J)
  280 CONTINUE
      WRITE(6,290)
  290 FORMAT(4X,'AT.%',5X,'WT.%',5X,'SU/100SV',5X,'A.',8X,'C.')
      WRITE(6,300)
  300 FORMAT(4X,'----',5X,'----',5X,'--------',5X,'--',8X,'--')
      DO 320 N=0,MS
      WP=(100.*N*W2/W1)/(100.+(W2/W1-1.)*N)
      AP=(100.*N)/(100.-N)
      WRITE(6,310)N,WP,AP,AA(N,0),CC(N,0)
  310 FORMAT(3X,I3,7X,F5.3,5X,F6.3,4X,F6.5,4X,F6.5)
  320 CONTINUE
C ---- LATTICE PARAMETERS VEA TEMPERITUR & COMPOSITION ----------------
  400 READ(5,*)A1,B1,C1,TO1,EV1
      READ(5,*)A2,B2,C2,TO2,EV2
      PK=1.38 E-23
      WRITE(6,410)
  410 FORMAT(4X,'TEMPERATURE',7X,'DA1',8X,'DA2'/)
      T=300.
      DO 430 J=0,20,1
      ALFA1=A1+B1*(T-TO1)+C1*(T-TO1)**2
      ALFA2=A2+B2*(T-TO2)+C2*(T-TO2)**2
      VC1=EXP((-1.6 E-19)*EV1/(PK*T))
      VC2=EXP((-1.6 E-19)*EV2/(PK*T))
      CALL LATTICE (R1,N1,AX1,S11)
      CALL LATTICE (R2,N2,AX2,S22)
      DA1(J)=(ALFA1*(T-300.)-VC1/3.)*AX1
      DA2(J)=(ALFA2*(T-300.)-VC2/3.)*AX2
      WRITE(6,420)T,DA1(J),DA2(J)
  420 FORMAT(4X,F8.2,5X,E10.4,2X,E10.4)
      T=T+50.
  430 CONTINUE
      DO 460 I=0,MS,MM
      DO 450 J=0,20,1
```

```
      C=FLOAT (I)/100.
      DA=DA1(J)*(1.-C)+DA2(J)*C
      IF(MM.EQ.5.OR.N1.EQ.1) GOTO 440
      CC(I,J)=CC(I,0)+DA
440   AA(I,J)=AA(I,0)+DA
450   CONTINUE
460   CONTINUE
      WRITE(6,*)'(a) LATTICE PARAMETER'
      WRITE(6,470)((AA(I,J),I=0,MS,MM),J=0,20)
470   FORMAT (' ',10F7.5)
      WRITE (6,*)'(C) LATTICE PARAMETER'
      WRITE(6,480)((CC(I,J),I=0,MS,MM),J=0,20)
480   FORMAT(' ',10F7.5)
500   STOP
      END
C ---- LATTICE SUBROUTINE -------------------------------------------
      SUBROUTINE LATTICE(R,K1,AA,CC)
      GOTO (21,22,23),K1
 21   AA=2.*SQRT(2.)*R
      GOTO 25
 22   AA=4.*R/SQRT(3.)
      GOTO 25
 23   AA=8.*R/SQRT(3.)
 25   CC=2.*R
      RETURN
      END
C ---- LATTIC SUBROUTINE --------------------------------------------
      SUBROUTINE ALLOY (B,A,J,L)
      DIMENSION B(0:J),A(0:J)
      DO 35 M=0,J,1
      GOTO (31,32,33),L
 31   A(M)=B(M)*SQRT(2.)
      GOTO 35
 32   A(M)=2.*B(M)/SQRT(3.)
      GOTO 35
 33   A(M)=4.*B(M)/SQRT(3.)
 35   CONTINUE
      RETURN
      END
```

APPENDIX B

Phase Diagrams

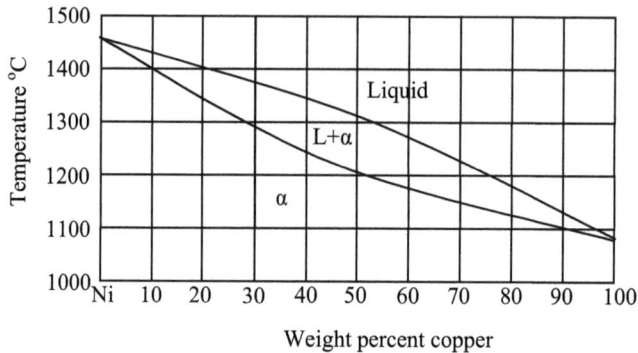

Appendix B1: Ni-Cu Phase diagram.

Ref. [7] Sidney H.avner, "Introduction to Physical Metallurgy", Second Edition, Tata McGraw-Hill Publishing Company Limited (1997).

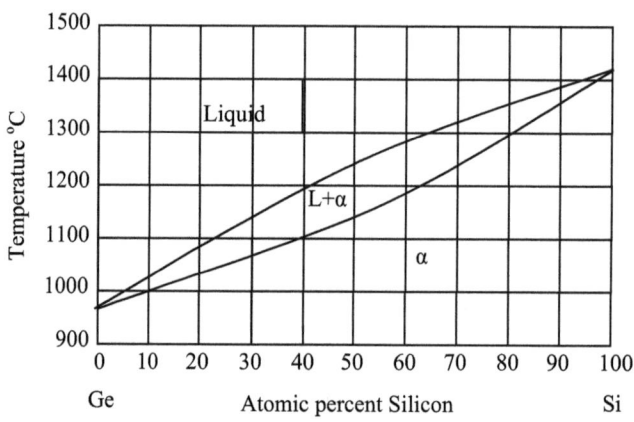

Appendix B2: Ge-Si Phase diagram.

Ref. [28] Ropert M.Brick, Robert B.Gordon, and Arthur Phillips, :Structure and properties of alloys", Third Edition, McGraw-Hill, Inc.(1966).

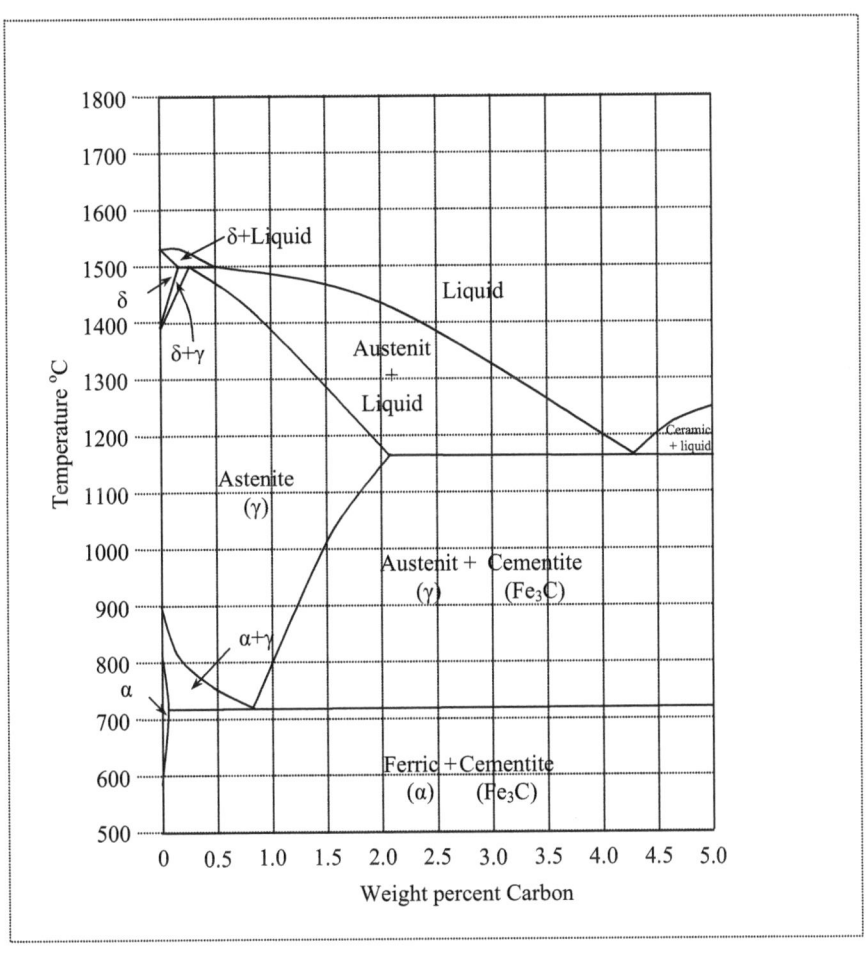

Appendix B3: Fe-C Phase diagram.

Ref. [1] Leonid V. Azaroff, "Introduction to Solids", McGraw-Hill, Inc, New York (1997).

APPENDIX C

Glossary of terms related to Physical metallurgy used in this study

Activation energy: The energy required for initiating a metallurgical reaction, for example, plastic flow, diffusion, chemical reaction.

Aging: In a metal or alloy, a change in properties that generally occurs slowly at room temperature and rapidly at higher temperatures.

Allotropy: The reversible phenomenon by which certain metals may exist in more than one crystal structure. If not reversible, the phenomenon is termed "polymorphism".

Alloy: A substance having metallic properties and being composed of tow or more chemical elements of which at least one is an elemental metal.

Alloy steel: Steel containing significant quantities of alloying elements (other than carbon and the commonly accepted amounts of manganese, silicon, sulfur, and phosphorus) added to effect changes in the mechanical or physical properties.

Alpha iron: The body centered cubic form of pure iron, stable below $910^\circ C$.

Atomic percent: The number of atoms of an element in a total of 100 representative atoms of a substance; often written *at.%*.

Austenite: a solid solution of one or more elements in face centered cubic iron. The solute is generally assumed to be carbon.

Bainite: A decomposition product of austenite consisting of an aggregate of ferrite and carbide. In general, it forms at temperatures lower than those where very fine pearlite form and higher than those where martensite begins to form on cooling.

Carbide:	A compound of carbon with one or more metallic elements.
Cementite:	A compound of iron and carbon, having the approximate chemical formula Fe_3C. It is characterized by an orthorhombic crystal structure.
Crystal:	A solid composed of atoms, ions, or molecules arranged in a pattern which is repetitive in three dimensions.
Defect:	A condition that impairs the usefulness of an object or of a part.
Distortion:	Any deviation from the desired shape or contour.
Equilibrium diagram:	A graphical representation of the temperature, pressure, and composition limits of phase fields in an alloy system as they exist under conditions of complete equilibrium. In metal systems, pressure is usually considered constant.
Ferrite:	A solid solution of one or more elements in body centered cubic iron. Unless otherwise designated, the solute is generally assumed to be carbon.
Gamma iron:	The face centered cubic form of pure iron, stable from 910°C to 1400°C.
Interstitial solid solution:	A solid solution in which the solute atoms occupy position within the lattice of the solvent.
Lattice parameter:	The length of the unit cell along one of its axes or edges, in a crystal: also called lattice constant.
Martensite:	Metastable body centered phase of iron supersaturated with carbon; produced from austenite by a shear transformation during quenching.
Matrix:	The principal phase or aggregate in which another constituent is embedded.
Metal:	Materials consisting primarily of elements that release part of their valence electrons. Characterized by good conduction of heat and electricity.

Phase diagram:	Same as equilibrium diagram.
Quenching:	Cooling accelerated by immersion in agitated water, oil or liquid nitrogen.
Recrystallization:	The change from one crystal structure to another, as occurs on heating or cooling through a critical temperature.
Solid solution:	A singe solid homogeneous crystalline phase containing two or more chemical species.
Solute:	The component of either a liquid or solid solution that is present to a lesser or minor extent; the component that is dissolved in solvent.
Solvent:	The component of either a liquid or solid solution that is present to a greater or major extent; the component that dissolves the solute.
Steel:	Iron base alloys, commonly containing carbon. In practice, the carbon can all be dissolved by heat treatment; hence, $<2\ wt.\%$ C.
Substitution solid solution:	A solid alloy in which the solute atoms are located at some of the lattice points of the solvent, the distribution being random.
Thermal expansion:	Expansion caused by increased atomic vibrations due to increased thermal energy.
Vacancy:	A type of lattice imperfection in which an individual atom site is temporarily unoccupied.
Vegard's law:	Calls for a linear variation of the lattice parameter, as a function of atomic concentration, in substitutional ionic solid solution between two component of similar structure.